Technology, Man, and Progress

William F. Cottrell

Miami University

Charles E. Merrill Publishing Company
A Bell & Howell Company
Columbus, Ohio

Merrill Social Problems Series

Under the editorship of

Edwin M. Lemert
University of California at Davis

Copyright © 1972 by Charles E. Merrill Publishing Company, Columbus, Ohio. All rights reserved. No part of this book may be reproduced in any form, electronic or mechanical, including photocopy, recording, or any information storage and retrieval system without permission in writing from the publisher.

ISBN: 0-675-09335-X

Library of Congress Catalog Card Number: 70-170880

1 2 3 4 5 6—76 75 74 73 72

Acknowledgments

The author and publisher gratefully acknowledge the following for permission to reprint:

Houghton Mifflin Company for "Technology and Social Change on American Railroads," from *Explorations in Social Change,* ed. Zollschan and Hirsch.

University of Chicago Press for "Technology and Societal Basis of Aging," from *Handbook of Social Gerontology,* ed. Clark Tibbitts. Copyright © 1960 by the University of Chicago Press.

Printed in the United States of America

Contents

Preface

Professor Cottrell undoubtedly knows more about the sociology of railroads and railroaders than any other man, perhaps more than any other man is likely to know. His book *Railroader,* his articles "Of Time and the Railroader" and "Death by Dieselization: a Case Study in the Reaction to Technological Change," many times reprinted, have awakened social scientists to the importance of technology of the railroad and its impact upon community and society. His major work, *Energy and Society,* referred to by a distinguished former student as a "classic theoretical study, wide in scope and with the force of mature scholarship," grew from the roots of Cottrell's conviction that railroading is the prototype of modern industrialism.

Cottrell's genius owes much to the continuity between his early life as a working railroader and that later as a scholar and teacher. Literally he is a railroader turned sociologist, having been born into a railroader family, raised in the shadow of the roundhouse, and having absorbed the skills and lore of railroading in many of its jobs: laborer, machinist, carman, trainmaster clerk, carpenter, strawboss, and even telegraph lineman. His early academic interests in transportation acquainted him with Cooley's works and thence led him into sociology as well as political science and economics. While no Renaissance man, Cottrell is truly an interdisciplinary man, a social scientist in the best sense of that term.

Cottrell's cross-disciplinary perspective was furthered by his early joint appointment in sociology and government at Miami University, Oxford, Ohio, and as he puts it, "by the need to keep up with [his] students in several fields." While his position and school precluded the training of graduate students nevertheless the number of undergraduates which he and his well-known colleague, Read Bain, inspired to go on to obtain Ph.D.'s in sociology has been phenomenal. Some of these have achieved considerable eminence. Cottrell's compelling teaching style and his ability to "turn on" students must be attributed to his special qualities as a human being and to the great range of his interests and knowledge.

The papers in the present volume reflect not only the breadth of his interests but also their applicability to pressing issues of the present day: obstacles to progress on our "sick" railroads, ecological realities, needs of our neglected aging population, and above all, the development of a science of peace to lead us away from the spectre of all destructive war. The thesis from which Cottrell never departs far emphasizes man's capacity to make choices necessary to conserve or further crucial values through feedback gained not only from cultural and social experience but also from direct experience with costs created by changing technology and ecology.

Cottrell is not a theorist in the grand style; his formulations belong to "middle range." Yet in another sense he has proposed answers to large questions about the nature and interrelationships of culture, learning, evaluation, technology, and the physical world. His conclusions on substantive questions are never far removed from data. This is well demonstrated in the paper on Caliente, especially written for inclusion in this volume, which revises some of his earlier ideas as well as challenging some commonly held ideas about socioeconomic exchange. Publication of the essay on peace, for the first time in the United States, epitomizes Cottrell's views that research, however contrary its findings may be to cherished policies, is vital to the survival of those values which make our society free.

Edwin M. Lemert

Part One

Technology

Chapter I

Technological Progress and Evolutionary Theory

A good deal of today's revival of interest in evolutionary theory seems to stem from the centennial of the publication of Darwin's great work. I am, however, enough of a student of the Sociology of Knowledge to doubt that this is sufficient to account for the whole of what is happening. In fact, the reasons why the theory got such widespread notice at the time it was originally published may give us some insights into the present revival. At that time traditional knowledge was cracking under the impact of early scientific discoveries and the application of some of that knowledge which was simultaneously giving rise to new social conditions. But traditional knowledge, technology, theology and morality were so closely wed that it was very difficult for either technological changes or shift in the social structure of the society to take full hold in the absence of some defensible rationale that would explain man's relation to his Creator. The whole static model which justified things as they were could be successfully defended if only men believed that they had been created as they now found themselves to be and were rewarded accordingly. Darwin showed how the Creator might have operated—not in a single act, forever establishing

3

what was to be, but in a series of acts taking place over time. His work was seized upon by those who stood to benefit from such a point of view with an avidity that was exceeded only by the bitterness with which it was attacked by those who relied for their security upon the preservation of things as they were.

In the Darwinian theory and its aftermath the competitive model developed by the Classical Economists was given both moral and divine sanction. Darwin's work demonstrated that the whole universe moved toward the ultimate realization of God's purposes through the survival of the fittest. To that of the Protestant Ethic was added the sanction of Science! Ruthless business and imperialism were thus scientifically demonstrated to be necessary and justified. It was only when the beneficiaries of the new age of scientific and technological progress had achieved positions of power and authority, and with it control over many of the rewards available for men of knowledge, that this competitive model became distasteful. The theory of evolution seemed to suggest that there is still the necessity for struggle, for invention and creation. But such an idea is a threat to those now thoroughly ensconced in positions of great power. It is more comfortable to adopt conceptions of a static world, where those who have engaged in struggle can pause to enjoy its fruits. Evolution was relegated into the background while less dangerous ideas, largely involving static models, became popular. The idea of progress that had, during the revolutions in the West, provided a universal symbol which justified striving, was dropped. Adaptation became the mode; each little segment was treated as a world in itself, developing local equilibrium through countervailing power.

But in parts of the world where men are not so comfortable, the model has not been dropped. Marxism supplies for many of them the ideological correlate to Darwinism. They believe that through struggle it is possible to reach higher forms of existence, and that nature ordains that through the achievement of superior science and technology Communism will provide the next superior form of man and society.

Our present concern is, of course, involved in this struggle over ideas. Is there something inevitable about the "progress" of the "underdeveloped" areas? Are there natural laws which will propel them willy-nilly into new forms of adaptation? If so, are these the laws discovered by Malthus, Smith, and Darwin and refined by the theorists of the West? Or is there something about Marx's formulations and those of his followers that effectively predicts that which must result from the interplay between what man discovers about the Universe and the way he uses it more fully to achieve his goals? Or is there still another set of requirements to be found in the Truths We Hold to Be Self Evident which will direct all men to seek what we seek, through the means we have used?

Only a little while ago questions like these were answered with extreme confidence. Science and technology can advance, we said, only along the lines and through the social forms developed in the West. Others must, to achieve our technical competence, pass through the stages we did, and once having done so they will inevitably become free; by which we meant they would internalize the values of Christianity, Capitalism, and Democracy and be able to achieve what their norms dictate. Today we are less sure that this is so, and even though our complacency has cost us dearly in many ways, we still do not like to look objectively at what has been happening in order to see why our predictions continue to be so far off the mark.

But in spite of our reluctance we have been forced to accept a revival of the idea that the world is a harsh place in which to live, one where force must be used to meet force, guile to meet guile, in the struggle for survival. We may not be so sure that the outcome of the struggle must be Progress but most of us are prepared to fight for existence, even in a worse world than this one.

So we in the Social Sciences are again looking at evolution as a model which has proved to be extremely valuable in the other natural sciences. But we are hardly equipped to use that model. Most of our research has been empirical and the bulk of it is based on observations taken in extremely limited bits of time and space. The means by which we manipulate our data are suited to deal with this kind of fact. Static and short run studies have had far more attention than historically oriented ones. The minutia which make up the bulk of our concern are selected to provide the base for such short run predictions. The business world, which has so largely controlled the development of technology, is interested in rates of obsolescence, styling, morale, limited movements of population, and so on. The educator has become extremely occupied with short run problems; the way to deal with the population bulge, the urban spread, and how to avoid teaching skills and knowledge that will be obsolete before they are mastered; even government is seldom able to support long range projects. There has been but little research money or brains to deal scientifically with the longer run social consequences of the technological changes that so often are at base the source of our immediate problems. Thus, even if it should appear desirable to use the evolutionary model, we can expect few significant findings from it in the near future. We will first have to raise a generation of students interested in, and equipped to investigate, a lot of things that have not hitherto been examined scientifically. In the meantime we have to live and deal with technological and scientific change and their social consequences by using ideas of either very doubtful worth or proved ineffectiveness.

Something of the difficulty we face is found at the very outset when we try to define what we mean by evolution. For some it is just history, nothing else. For others it means only change. One student looks upon it as the expression of Newton's Second Law of thermodynamics writ into varied forms. Evolution implies, for another set of observers, movement toward higher forms, while still another identifies it with more complex ones and yet others take the position that it is the way the greatest possible population of a given type can be maintained in a given set of circumstances. Obviously the definition chosen represents as much what the given investigator is prepared or willing to study as it does about "what is out there" to be studied.

It is on this kind of heuristic basis that I have chosen the definitions I will use here. I am not trying to synthesize or provide the greatest common denominator of other definitions. What I think of as "evolution" is a concept which implies that we are making or have made an analysis of the relationships which extend through time among a cluster of patterns to which we give our attention. We want to know what recurs and what differs during a given period; which occurs first in the system of coordinates of space in which our clock is located, and which follows. I give the name "pattern" to the complex or configuration whose recurrence or disappearance we are recording because most of us in our culture seem to make abstractions easiest in terms of geometric form. Whether this be a consequence of the fact that we communicate most of the time through arbitrary arrangements on a plane surface, or for whatever other reason, I am not prepared to say. Nor am I arguing that some other artistic form, some mathematical expression or other kind of symbol might not serve equally well. The chief point is that while events themselves have multitudinous connections with each other, extending both in time and space, the symbols which stand for some of the common characteristics of these events represent only a few of these relationships. It is only by neglecting a great many differences between specific occurrences that we can classify them as being the same thing. By "pattern" I mean to imply an arbitrary selection of this kind. To come back then to the line of thought we were following: Having chosen a number of criteria, observable with the instruments we are using, we find that these criteria assume patterns that stand in more or less persistent relationship one with another. If we are to find the conditions under which these patterns of relationships change or persist, we are forced to take samples at time intervals. When the sample has characteristics that fall outside the range of what we are willing to call "the same thing" we record change. On the basis of such samples we discover that the patterns that follow one another in time exhibit both persistence and difference. Where persistence follows persistence in all the patterns under observation there is no way that we can predict the probability of change. When some patterns change and others persist we assume that the connections between them are not revealed in such a way that they help us to increase the accuracy with which we can state the probability of change or persistence. It is only when observed changes in some patterns are regularly followed by particular kinds of changes in others that we have increased our ability to state more accurately whether change or persistence is likely, in what direction, and of what character. We look primarily at the succession of events from the past into the future rather than the reverse because the knowledge we are looking for is most useful in dealing with the future. It will be clear by this time that my definition of evolution is only a special case included in a definition of causality. "Evolution," as we shall use it is an effort to understand one series of cause and effect relationships.

Like all science it begins with the observation of specific events, but it does not stop when this history has been recorded. If we are to use it for scientific prediction evolution cannot be synonymous with history or with change. Whether or not in a given case it is useful to include the evolution of atoms and stars in the effort to understand the evolution of machines and patterns of social relationships is a heuristic problem. All we ask is "Can we, by adding the history of this pattern during the period of observation we are using, to that of the other sets of patterns also under observation, increase the accuracy with which we can state the probability of the specific kinds of change in which we are interested?" And in the interests of economy, which is important in assessing the heuristic value of any course of action, we must also ask "Is the gain in accuracy worth the cost of the proposed action?"

This has been perhaps a rather long-winded kind of justification for my definition. I have given it only because in reviewing the literature I find that a great deal of the time and effort wasted in controversy stems from a failure to lay out the premises from which the argument was made.

Darwin was concerned with the origin of the species. If we confine the use of the word evolution to efforts to solve the problem he was trying to solve we will of course have shorn off most of the substance of our present concern. But if, on the other hand, we define evolution to mean all of the causal connections that extend among those classes of things we call technological, physical, or cultural and those we call biological, we have undertaken an impossible task. This forces us back into a heuristic selection. We have to hunt through the research on how man and other organisms changed to see whether we can find knowledge useful in predicting the direction that technological change is likely to take, something of its magnitude, the effects it has had, and perhaps its specific applicability to the kind of things that go on in underdeveloped areas. In so doing we must perforce neglect a very large part of the evidence collected for other purposes by those interested in evolution for other reasons. But we will not be looking for mere analogies and parallels. The findings of evolutionists have been put into symbols that represent only a small part of what others, differently oriented and using different instruments, might have observed or did observe in the situations symbolized. Technological change was a part of, not apart from, organic and physical change also being observed. Put another way, we will not treat technological or social events as being merely analogous to biological or physical events—they are simply another among the many patterns of relationships different observers might discover among the same events.

Darwin did not have to give much attention to the class of observations that we call technological. He drew most of his evidence from a period of time in which little or nothing that we commonly call technology had yet come into existence. He could safely give limited regard to the special influence which such things as birds' nests, spider webs, beaver dams, bee hives, ant hills, and even pebbles nudged into position to block the entry into fishes' nesting places must have had on the appearance and survival of species. In a sense the use of such artifacts as these deserve to be called technological, and a thorough review of their impact on the evolution of certain biological forms might demonstrate that influence to be considerable. But certainly it would add little or nothing of concern to us. However, the habit of assuming that biological change could be understood completely in terms of physical, chemical, or biological changes in the natural environment unmodified by man was a dangerous one, once the origin of *man* became the subject of inquiry.

What in fact resulted from this outlook was this. Since physical and chemical patterns as they occur in nature can be considered to be almost constant during the time *homo sapiens* has been evolving, it must have been biological change that accounts for what he is and what he has done. Technological and social change merely represent the working out of the prior biological development. A generation or more was spent digging up the evidence to prove this point and all kinds of social policy was "justified" by the assumption that it was correct. The "fittest" were endowed by their biological superiority with the right to rule others, lesser breeds, who should be thankful "that God had in his infinite wisdom provided more able and intelligent men to govern them and prevent their being victims of their own mistakes". Of course it would be absurd to say that imperialism and colonialism were "caused" by the error in interpretation of evolutionists. There might have been other myths to rationalize and influence what was happening. The point is that this was the myth and it was bolstered for a time by what was regarded as final scientific proof.

We now know, of course, that the kind of biological organism that survived to become *homo sapiens* very probably would not have done so had there not been previous advances made in *technology* by that forbear or some of his fellows.

I suppose it will be necessary at this point to define what we mean by technology. Broadly it could be made to mean the same thing as technique, the regular way of getting things done in a society. For reasons we will explore later this is too wide a definition of technology for us to use here. But some such definition did fit the categories that the Darwinists made. For their purpose the dichotomy, heredity-environment, was necessary. Certain kinds of persistent patterns could be shown to pass biologically from generation to generation. These represent one category, heredity. The other category, environment, might consist of other kinds of recurring patterns or of irregularities. The important thing to know was that if patterns persisted they were not being preserved by biological heredity.

Regularities in environment that persisted beyond the life of a generation quickly became apparent. These were said to be due to the direct effects of the physical and biological world. Later it was found that the patterns observed operated within much narrower limits than those imposed by "nature." Side by side, in what the geographers called the same environment, were persistent patterns differing from one another. Both their differentiation and their persistence required explanation. Their persistence was found to lie in the fact that there was a kind of heredity not carried on through events taking place within the bodies of organisms. To such patterns were given the name culture. Since through the observation of culture it was possible greatly to increase the accuracy with which they could predict the probability of relationships in which they were interested, the evolutionists expanded their model to include not only biologically transmitted patterns but also persistent patterns not so transmitted. It is important for us to remember this. Whatever kinds of nonbiological factors that had the effect of producing patterns persisting through generations were for these evolutionists one category, culture.

When anthropologists got around to examining closely the actual process through which the perpetuation of these patterns was secured, they found it to be more complex than that involved in genetic transmission. Conception takes place at a single moment in time. At that moment the whole genetic endowment from the past is transferred to the organism that will be patterned by it. Obviously this is not the case with culture which has to be learned during a considerable period under quite variable conditions, from a number of sources. It may be dangerous to neglect the consequences of these differences in our effort to explain the likenesses. The great bulk of the patterns are passed on at least in part through symbols, and some anthropologists would confine the word "culture" to what is symbolically transmitted. Some patterns are transmitted through a combination of experiences. One learns to operate a lathe or ride a bicycle the way he learns to swim, by doing it. The physical objects, modified for man's use, that were passed from generation to generation contributed continuity according to their durability. Skills and judgment involved in handling these artifacts are learned, in part, from symbols, but in part learning takes place by direct feedback extending from the subject to the object and back again. It will not do to disregard this process and rely for our understanding of either change or persistence purely on the effects of symbolization—or symboling if we perhaps more accurately use the word as a verb to indicate the active process.

The scientist in field or laboratory learns from the reactions of materials directly taken from nature as well as from the symbols provided by his culture. What he has learned cannot correctly be imputed to come solely from either source. Similarly, the experiences of those who work with artifacts give them a different meaning than would have been conveyed were symbols alone involved. The consequence of this difference in the learning process was noted by the anthropologists. Students learned that some aspects of a culture could be more clearly understood, and changes in it more accurately predicted, by studying physical and chemical facts than by observing symbols alone, while in the case of other aspects this was not the case. To the first the name "material culture" was given, and the residue was called "non-material culture." For some theorists technology is identical with material culture.

This leads to confusion. In the functional sense, in that they provide the means to carry patterns that extend from generation to generation, the categories should be treated together. The symbols that deal with material things are transmitted exactly the same way that symbols dealing with other things are transmitted. If, however, culture is only that which is symbolically transmitted, then that which is learned from artifacts through direct feedback is not culture at all.

We can't hope to undo the confusion that exists as a result of this situation. But what we want to make clear here is that "technology," as we shall use it, includes traditional knowledge gained by earlier men, including a lot of prescientific propositions developed by them and transmitted symbolically. It also includes skills and judgment learned on the job during the lifetime of those presently working. In addition there is knowledge newly invented, like science, which was not passed on to this generation but invented by it. For us, here, technology is not synonymous with material culture, nor with non-material culture, nor with culture, because it cannot be most effectively understood by trying to deal with it as if it were.

There has generally been an effort to identify technology with knowledge about and skills in using tools and machines. This could easily be accepted as our definition. But if we use it that way we must understand what this does in terms of evolutionary categories. If it is difficult to deal with an object as if it were physical or cultural, when it is both, man's control over chemical processes provides an equally confusing kind of relationship to deal with. Man's control over fire, which apparently began as far back as the early Pleistocene, has had a good deal to do with his evolution but can we call it a tool? Fire could be used for warmth, permitting man's forbears to survive in climatic zones where they otherwise could not have done so. It was also itself a weapon protecting him against predators. It could be used to cook food, making it possible for him to use for his survival plants that he could not otherwise digest. It may have been used to harden spears and to shape wooden implements at a time about which we can learn almost nothing. But more significantly for evolution it was a means to alter the whole face of a region with fire. Large areas of plant and associated animal life that otherwise could have become dominant were destroyed. This could be done repeatedly so that even the "long run" development that otherwise might have taken place never did. Similarly, certain types of plants that require considerable heat to permit them to germinate got, through man's use of fire, life chances that otherwise could not have existed, and so affected the life chances of other plants and animals.

At least at some places on earth the use of fire should be considered as a factor to be taken into account as much or more in explaining the survival of species and associated organization than the use of speech and tools, the other two human possessions which often are used to explain why an ape-like hominid could evolve into *homo sapiens*. Shall we then place the use of fire in the same classification with symbol and speech, with tool and learning-by-doing, or give it a special classification? Certainly its effects on survival are great and extend far beyond our capacity to grasp, influencing man's development in ways we know not.

For our concern here it may not matter how we classify it, but at least we ought to be put on warning that when we exclude the use of fire from our analysis of technology, we exclude from consideration a significant variable that might, if we could get evidence, greatly increase the accuracy with which we reconstruct man's history. Fire was probably used before tools were invented. But we have no way to know that. To reconstruct the past we have had to rely on stone and fossil. By its own nature fire destroyed in most cases the evidence of its use. So we cannot correlate evidence of changes in the use or extent of the use of fire with evidence of biological change. And so we are barred from using it directly. It is when we begin to make more derivative inferences from the evidence we do have that we need to exercise caution. From a knowledge of artifacts and fossils we have established certain kinds of ideas that assume that the change in tools must have caused changes in the organism. It may well be that in many instances it was the controlled use of fire in addition to or instead of the use of tools that gave rise to biological change. Perhaps some of the conclusions that we have come to with reference to the kind of social organization that must have existed at a particular time and place rest too exclusively on the idea that it was only tools, changes in symbols, or organization that gave man advantage over competing organisms.

Whether or not we include knowledge about and control over the fire as part of technology, we must include knowledge and skills connected with the use of tools and machines. So taking this as the core of our concern, let us take a look at the way technology affected and was affected by the evolution of biological types. The work of the archaeologists has established beyond doubt that changes in the form of tools and other artifacts are associated with changes in the distribution of fossils of various types dating from the same or succeeding periods. Man through changing his environment changed the conditions under which selection among types of man or his predecessors took place. Changes in environment do give rise to changes in heredity, if not directly in the old crude Lamarckian sense, indirectly, through affecting the statistical distribution of mutant types in the gene pool. Cultural invention and innovation do not necessarily give rise to mutation (though the invention of radioactive materials certainly does) but they often affect mightily who will breed and survive and so what genes will disappear from the pool. It is no longer possible to say that man became what he is by "allowing nature to take its course" to some foreordained end. We have in fact reached the point where the kind of man he was is often identified by the artifacts he left behind and his characteristics are inferred from them. What must be emphasized here is that while the scientific base for the idea of evolution through biological superiority has been pretty well destroyed, a great many elements in our culture are still what they were when it was held to be valid, and some of them are basic to our present attitudes about underdeveloped areas.

Akin to this belief in a kind of biologically natural superiority is another that denies any place to values in human evolution. By value I mean "the factors that affect choice." The early determinists, caught up as they were in a struggle with metaphysicians, theologists, and vitalists felt that they had to deny the existence of choice, lest they let supernaturalism back into their considerations. They insisted that they had found in chemistry and physics an invariate order. A caused B and that was that. Today of course it is apparent that there are a terrific number of irregularities yet to be accounted for in these sciences. Once the irregularities appear however, there is an immediate effort to create means of observation that will permit the discovery of new patterns, and if this is successful a new model which takes into account the newly observed regularities becomes heuristically desirable.

If, with the use of a model that includes means of observing the influence of choice, we can more accurately predict survival than we can using one that neglects this factor, then we are bound by heuristic imperative to use it unless the cost is greater than the gain. Once we have established the fact of choice we can proceed to examine the factors that affect choice and perhaps move still another step closer to accurate prediction in those fields in which choice matters.

Peculiarly enough there was little criticism of Darwin's use of the concept of choice in his demonstration of the way evolution takes place. His idea that sexual selection was responsible for fixing some of the characteristics of birds was crude and needed refinement, but it is not possible to dispense with it entirely. Nobody denies that when a bird chooses one rather than another as mate from among those available he affects the gene pool and that if there are large numbers of individuals carrying some specific types of genetically determined characteristics who are regularly excluded from participation in breeding the result will be to alter heredity.

The mechanism of choice is not invariate. If the single-minded pursuit of the queen bee by all the drones represents "instinct" then something else must be held to account for the fact that a particular pair of geese mate and subsequently disregard the availability of what seem to us at least to represent sex objects no different from the one chosen. Deer choose browse over hay when browse and hay are both available, but any stock raiser will testify that they will eat hay when no browse can be had. Such facts as these and thousands more that evidence preferences among animals lower than man are accepted without argument. Nor can any one doubt that in turn the choice by an animal of one plant as food rather than another affects the ecological balance in the habitat of that animal.

The mechanical model of cause and effect that has been in use in our culture so long made it very difficult to take teleology into account. The development of the theory of servo mechanisms has given us a new means to understand some of what takes place in man's behavior without forcing us to leave what we regard as being scientific ground. Although it has long been possible to observe such things as a hawk directing his flight not toward the point where the bird was first sighted, but on a course that anticipated where he would be at the moment of collision, we nevertheless avoided explanations that involved goal-directed behavior. But now that we can design a missile that will hunt a missile perhaps we can also accept the purposeful human behavior that invented it. Through the use of feedback theory we have come into possession of a model that may contribute to our understanding of processes that are going on.

It is pretty obvious that the use of a model that permits analysis of preference is heuristically justified in the study of evolution. With its use we can set about to discover the factors that affect choice using generically the same kind of process that is involved in other kinds of scientific endeavor. If anybody objects to my use of the term "value" to indicate the factors that affect choice, and the use of the term "evaluation" to indicate the process by which preference is arrived at, he is free to assign any other words that to him seem to be more fitting.

Coming back to the analysis of evolution, it appears that man is the product of evaluation as well as variation and selection by non-teleological natural forces. The consequences of choice may be such as to alter future alternatives in a way to make such choice again impossible. It may be such as to lead to the elimination of the organism choosing, or to that of some other organisms sharing the environment. It may also, through positive feedback, reinforce choice so that its future probability is increased. A great deal of culture represents the results of just such feedback. But man is led to choose not only by experience in his own lifetime but by that of his forbears on the basis of the results apparent to them.

The kind of nervous system that permitted learning in the early years so that culture could be transmitted to succeeding generations had obvious advantages for survival. Those forbears that could use tools increased their chances of survival over those who could not. So also those who had accumulated experiences, such that they need not make the mistakes others not so equipped were more likely to make, could, through the exercise of choice, modify their environment and that of future generations. The use of tools and fire and culture, and the exercise of choice, mutually affected what took place. To give priority to tools over values, biological type over culture, culture over values, or any other kind of causal assignment, on the basis of the record, is arbitrary. Imputation of causation is a function not only of the observed but the observer. If one chooses a specific point in time when a particular change is taking place in one or another aspect of the complex he can show subsequent change in other elements, but often, had another arbitrary starting point been chosen, change could have been shown to arise in the reverse order. If to get an "objective" starting point from which to impute final cause we keep going on a course, each step of which leaves out much of what in fact took place and was involved in what followed it, we continually know less, not more, of what the future will look like.

Perhaps because of the apparent futility of such a regression, perhaps because of changing concern and motivation, perhaps because of the greater availability of means, or for other reasons, those who studied organisms in relation to their environment increasingly devoted themselves to the analysis of single configurations extending through a short space of time.

This type of ecology has provided the means to discover many of the intimate interrelationships that pervade the ecological complex. It has permitted us to learn much about how, under a given set of conditions, a given plant or animal population will sort itself or be sorted. Some general propositions have been substantiated well enough that we can say that, neglecting the impossibility of mutation of invasion which will introduce organisms not now to be found here, taking into account only predictable changes in soil, climate and topography, "this kind of stable equilibrium will appear among organisms in this habitat."

Such a model has great heuristic value for those studying the impact of man for he can be treated, at least at the moment of his invasion into a habitat, as an independent variable whose presence gives rise to new hitherto improbable patterns. Man's influence in changing the face of the earth can by this process be to some degree inferred. The record shows it to be far more considerable than it was once thought to be. But again he is not apart from but a part of the ecological complex, and whether he changes the face of nature because he is the kind of animal he is, because of the culture, tools, or organization he possesses, cannot be directly known. Only where differing sets of men, using different tools, different cultures, different value systems, have invaded similar habitats can we make this inference.

In spite of a multiplicity of cross cultural comparisons there is no general agreement as to the influence of various elements of the complex. What we have come to agree upon is that the animal-man can be treated as being the same, whatever the habitat. We no longer can expect that isolation will be maintained over a long enough period of time that mutations will be differentially selected out and separate gene pools established. From now on one of our previous variables can be treated as a constant.

But if this is so, and I know of no fact more generally accepted among anthropologists, then at least part of the evolutionary model loses direct interest for us here. No matter how he got here, or what influenced the character of his heredity, we have to deal with man's gene pool as an unchanging fact during the last twenty-five thousand years or so. It is possible that there are some changes taking place in it that will become apparent to observers able to make use of a longer period of observation than we have, but they cannot be established now. Man is still one species. What we discover about the influence of technology upon man's development during the existence of *homo sapiens* is not recorded for future generations by patterning DNA. We will have to look elsewhere if we want to know about changes in the kinds of societies he builds or the psychological make-up he exhibits. It is also now possible to rule out certain kinds of patterns as being inherited. We do not have to change the gene pool either to increase or decrease this prevalence. It is also now apparent that a great deal of what we thought to be self-evident attributes of all human beings turns out to be only characteristic of a few people occupying a limited space, for only a limited time.

As a matter of fact, of course, our concern is primarily social and cultural. It gives us no comfort to know that no matter whether we regress to fairly simple hoe cultures or progress to fulfill all Madison Avenue's fantasies, man will still (barring mutations whose selection may change it) share the same gene pool a hundred years from now that he presently does. What we are here primarily concerned with is the persistence or change in certain kinds of patterns we label technological, social, political, economic, or religious.

Problems connected with the identification and the discovery of such patterns, the creation of effective typology, and analysis of their covariance with other kinds of patterns remain very difficult. One reason for this is to be found in the nature of their transmission from one period to another. As we pointed out earlier, in biology we now know enough that we can clearly separate in our thinking the elements involved in change that derive from alterations in the gene pool from those that take place otherwise. While it is true that the nature of the species is inferred from the statistical description of a host of individuals, it is also true that each one of these organisms is endowed with all of the potential it receives genetically at one moment. A man's biological inheritance results from a single act by his parents. This is not true as it relates to his social inheritance. A long series of experiences is required to enculturate an individual. In no society is anybody endowed with anywhere near all that can be found by adding together all that is culturally transmitted by all of the individuals that make it up. What part each gets makes a difference in his behavior. Nor is there any period in the life of the individual that we can say he has received all from the past that he will ever receive. During the period that he is having symbolic experience with things from the past he is also having direct experiences, many of which are not culturally defined for him, and he may well react to such experiences in ways that modify efforts to induce him to act in culturally sanctioned ways. The deviation from cultural norms increases with the complexity of what must be learned, the order in which it is learned, and the significance attached to learning it. Those who study culture patterns find it harder to categorize them and order their appearance in such a way that cause and effect relationships among them can be imputed. Perhaps the most significant difference between the model that helped biological evolutionists see what happened and the one necessary to understand cultural evolution is the place that must be given to purposive action. A great deal of biological evolution can be dealt with effectively without taking this into account. The mechanism of choice in a great many cases is "instinct" that is transmitted biologically. Moreover, since selection is the primary process operating, the origin of non-genetic innovation is not as important as its consequences. The error involved in neglecting this source of variability is not very great. However, as we have seen, once the origin of man became involved his purposive behavior became increasingly important, and the shape of things to come often could be predicted much better if this factor were taken into account than without it. Models designed to predict cultural evolution must, if possible, discuss the bases for cultural innovation as well as the selective factors involved in their survival and that of the population in which they take place.

It is apparent that evolutionists have been concerned with the numbers of various genetic types that survive in a habitat. For many of them the study of evolution consists of an effort primarily to predict more accurately the size of each population occupying a habitat after a given period. For them, the significant fact about man is that he has increased in numbers so dramatically and has come to occupy so many different kinds of habitat. Even the explanation of why this is so is a considerable task. But most of us want to know not only how many of us there are going to be but also something about where we are going. Are we necessarily driven to increase our numbers up to the maximum fixed by the food chain in the area in which we live? If so, what will happen to our cherished values and institutions? Does evolving technology require a world different than that envisaged by our values, or can we set up values and shape technology willy-nilly so that it serves them? Or are there moral principles with which men are "endowed by their Creator" which necessarily govern, overriding both biology and technology? And even assuming that there are no such ethical principles as this assertion about man's relation to his Creator, are there some others to be found in the nature of the organism itself? For example, can the kind of man that in his childhood responds to efforts of older men in a way that permits him to be enculturated exist except in societies that induce men to teach their children? Can such an organism learn except under the influence of personal interaction that includes love and affection? And can an organism of this kind ever be taught successfully to make the choices necessary to the survival of extensive organization and advanced technology if it robs humans of the continuing opportunity for these experiences? If not, then this elementary set of ethical principles becomes a basic fact. Any culture, any technology, that attempts to function in disregard of these propositions is as certain to be adversely selected as is one that disregards other imperatives of its existence.

If then there are tendencies in humans that induce them, like other organisms, to increase their numbers up to the physical and biological limits of their habitat, and if there are technological imperatives driving man to achieve the greatest possible control over energy, there are also factors involved in organization and in learning which must be taken into account if we are most accurately to predict emergent culture. In fact, of course, a very large part of the influence of habitat and technology operate by influencing choice as well as by selecting out those choices that produce adverse results. The model we use for the analysis of cultural evolution cannot dispense with the knowledge we have about how other organisms evolve, but it must demonstrate how what we have learned through its use and in other branches of science, interacts with human purposive action. Most anthropologists will agree that man shapes his culture, and on the basis of its effects alters or retains it. The point where many of them part company lies in their interpretation of this process. Certainly to the degree that teleology is involved, we must start with the human being, the chooser who acts in ways that perpetuate the pattern or alter it. In "making up his mind," "evaluating the situation," or however we term the act of choosing, the actor responds through feedback from his previous acts, or on the basis of cultural conditioning, anticipates what the response will be and acts as he would if the response had already taken place. But after he acts he gets a response that is a result of what happens in the situation. This may or may not correspond with his anticipation. The response may so alter that situation that further choice is impossible, like destroying the chooser, or so narrowly limit the range of his future behavior that his choice no longer affects the outcome. On the other hand, the discrepancy between anticipation and realization may still permit further choice. In these cases man can respond selectively to what happens. That is to say he may react as if only part of what happened did happen. This may be due to peculiarities of his own organic makeup (like, for example, being blind or deaf) to the idiosyncracies of his own learning, or to the kind of cultural patterning common to most of the humans in the area where he is operating. Culture will have, in any case, set up a perceptive base on which the adult human must function. The response he gets will be defined by the enculturated individual selectively and the pattern of his past will be re-enforced or altered on the basis of this selective response. If the cultural definition is sufficiently adequate, the response of those sharing it permits them to initiate new acts to which in turn they will respond selectively. This may go on for a very long time. The culture provides the individual with definitions of response (meanings?) that serve to keep people acting for generations in ways that those not so enculturated may find to be fantastic. Looking at such behavior the pure culturologist says "culture can make anything good" and cultural relativists deny that there is any standard common to mankind which affects his efforts. For him there are no goals except cultural goals. No values not purely a product of culture. In a sense this is true. If culture be treated as a residue, what is left after all the factors that operate to select cultures have had their effects, then of course what is found there includes all norms, the results of all strivings, as well as all knowledge that will be symbolically transmitted. But how much of what was there yesterday is still there? How much of that was selected out by the operation of non-cultural factors? How much of what is there now is new, now to be symbolically transmitted but not learned that way? We cannot find these things out by studying culture itself. If we confine ourselves to this kind of evidence, of course we will find that the only kind of sequences there are, in terms of which to impute cause and effect, are cultural sequences. But if, on the other hand, our model permits us to look elsewhere, we may see that culture change was preceded by technological invention, or that certain kinds of deleterious social relationships were selected out of that culture when new knowledge made it possible to discover their influence, or that changing biological facts, like the growth of population, made what had been tolerable cultural propositions intolerable. Perhaps it is only in a comparatively short run that "culture can make anything good." (That is to say that what culture calls "good" will persist.)

For example, a physiologist or epidemiologist looking at a culture might come to a different conclusion about the persistence of certain culture patterns than one locally enculturated. He would point to the physiological resultants of the behavior that took place without reference to their cultural definition. He would show that certain culturally-ordained practices in relation, for example, to diet and sanitation are regularly accompanied by certain morbidity and mortality rates, without reference to whether or not these practices are defined culturally as being good or bad. From this information, the future size of the population, and the level of its accomplishments, to the degree that they were dependent on physiological processes, could be predicted pretty well. In turn, an ecologist viewing the ability of this population to compete with others for sustenance might well be able to predict that even though the culture lost nothing of its ability psychologically to induce its carriers to make selective response necessary to preserve it, the population itself might in a relatively short time disappear, and with it the culture it was carrying. What we have been saying is that there is goal-oriented behavior, that goals may be provided by culture, and culturally sanctioned goals may be pursued for a long time that frustrate the primary impulses of the human organism, provide technologically inefficient means to the sought ends, and are physiologically debilitating. But a culture operates in a milieu which includes other populations, often sharing other cultures, plus biological and physical elements too. Its survival depends not only on its ability to perpetuate itself psychologically in a population. It must also fulfill demands made on it by these other factors. But we must be wary in extending to cultural survival the kinds of all-or-none principle that work with the survival of genes. A gene lost from the pool is forever lost. Whatever elements it contributed to the organic pattern will from that moment cease to exert influence on the future. But cultures are transmitted piecemeal, and there is continuous alteration going on in the society using it. Soon after two cultures are brought into contact diffusion takes place and each will soon be discovered to have borrowed from the other. What culturally is lost by the disappearance of one population may well survive in another. Natural selection operating on the biological base need not be directly accompanied by any similar or identical cultural change, and the reverse is also true. A simple way to treat this fact in the theory of cultural evolution is to assign to biological and physical phenomena a static role. We say that these factors set limits to man's behavior, and assign to culture the whole of the basis for dynamics. Culture, we say, produces culture. One culture may have impact on another, but it is not necessary dynamically to relate cultural change to anything not included within the cultural definition. Yet we readily admit that cultures seem to interact with habitat, that for example, Eskimo Culture is different from that of the Maori, and that each is suited to the environment.

If we return to the idea that technology evolves through feedback, we get a model that permits such a dynamic interpretation. What is learned from the dynamic responses of the habitat to culture-directed behavior gives rise to choices different from those that would occur when identical limits were imposed by habitat and only symbolic interaction between individual organisms were involved. Nature does not confine itself to saying "this far and no further." Its responses are often dynamic. They become different than those anticipated in the culture and so give rise to different experiences, on the part of those acting, than were anticipated. If such experience is widespread, new anticipations develop, are subsequently transmitted by symbol and so become part of the culture. What is today deviant behavior becomes tomorrow normal, not necessarily because there is some "strain toward consistency" in the culture itself, but also because feedback from nature is selective in dynamic terms. If men fail to respond differentially on the basis of nature's responses, their survival itself may be endangered. Take for example declining plant yield from a single plot of ground. Culture can ordain that man move about and, as a result, he avoids the consequences of this natural response. Or it can provide means to restore fertility to the land. But unless it provides means to stop declining rates of plant growth, it cannot ordain that men remain there permanently and carry on in each generation as they did the generation before. Man is not wholly rational, but we cannot deny to him an element of rationality. Being a sentient, goal-directed organism, he adjusts culture through feedback. The physical and biological worlds interact dynamically with the cultural world. In the process man alters the order of his choices, some acts seem now to be more significant than they were, others less so. Values, culture, and habitat make up a network of relationship that persists in dynamic, not static, equilibrium. Not all of the dynamism can be attributed to man.

Short run stability may be followed by rapid change. This often accompanies a mutation, an invasion, or an invention. Population which has been in stable equilibrium undergoes rapid change, but slowly it again comes into balance with the resources it can wrest from its environment. This kind of change over time provided the evidence from which our theories about evolution are derived. But recently we confront a situation in which change is far more rapid than it ever was, and continues over a longer time than did change that took place at any corresponding rate. The old idea that change could be stopped by the fact that men got tired of striving, became anxiety ridden and apathetic when change went on too long, or otherwise reached the place where they no longer willed change, will not fit the present facts.

Since nothing like it happened before, we will have to look at recent kinds of change if we are to find reasonable answers as to what is now happening in social and cultural evolution.

One theory of change assumes that what is now going on is due to the geometric accumulation of culture traits, particularly in the realm of "material culture." Certainly there is evidence that such a progression has taken place. But this fact is not self-explanatory. Why this sudden growth? Does it stem from a new kind of creativity in men? Are there an increasing number of geniuses being drawn from the gene pool? Or have we found a way to get dynamic responses from nature that men previously were unable to secure?

It seems to me that it is profitable for us to explore this latter hypothesis. Certainly a ubiquitous fact about the last five or six hundred years is the increasing mobility of men and their increasing possession of things. Both require increasing control over energy.

In the use of the energy concept we need to be careful lest we involve ourselves in tautology. Energy is a word used in the sciences to describe some regular relationships found to exist in nature. It is often also cited as being the cause of such relationships. In fact, of course, we have used the same symbol to point to two kinds of relationships among the same class of events. In spite of the confusion this generates, the energy concept has been extremely fruitful in other sciences, and we will use it here just as it is used there. We can measure energy by the effects it produces on our instruments at the moment it is converted from one pattern (or form) to another. Because this has been done, it is also possible for us now to state by formulae what the ratios between energy in two different forms are, and compute the amount of energy involved when the changes from one form to another take place.

We observe all of the patterns we observe through energy conversions. Some of these conversions take place within the body and others outside it. All of the patterns we have been talking about can, from one point of view, be treated as arrangements of energy. Persistent relationships—(structure)—and dynamic relationships that involve describing the flow of energy through structure make up the observable world. We could use a model of this kind to deal with patterns of value, culture, organisms and things. At the moment it does not appear that such a model would permit us to predict probability more accurately than we can now do. But certainly we are required heuristically to investigate and see where it gives better results than those we now use. We are increasingly equipped to use it by the work being done in other sciences. The present concern of physicists with energy is too ubiquitous to mention. Organic chemists are increasingly using it to see how energy conversions are involved in rearrangements within the molecule. We know enough about photosynthesis to begin to unravel the energy relationships involved in synthesis and organic growth. Through the study of metabolism physiologists are beginning to understand how energy conversions take place in the body. It is now possible to trace energy from the sun—the source from which, with negligible exceptions, it comes—through the organic community to entropy. Some evolutionists have begun to evolve a hypothesis and build an energy model to study organic evolution. If it works out, it will bring together in a larger framework some scattered knowledge we already have. Most organisms taken as physical units are steady state systems. They grow to a point where energy inputs equal outputs and stop there. But taken as a class they represent expanding systems. That is, they reproduce in such ways that the only limit on growth is found in external conditions. They may be limited by inroads made on them by other organisms, or by the energy that they can control, but they are not self-limited. Organisms "capture" the radiant energy of the sun as a means to perpetuate the patterns that differentiate them one from another. Various gene patterns carry on this process of "capture" by different means. Complexes of genes result in the formation of organisms that constitute a food chain, each organism being dependent for its own supply of energy from forms created by another, performing a different operation involved in its own survival. A particular organism then is dependent for its energy control on earlier patterning by others as well as upon its own capacity to control the conversion of energy. In a particular habitat there are a number of possible combinations of such organisms. These combinations vary in pattern as well as in numbers. Together they control the conversion of part of the radiant energy of the sun that reaches them in their habitat. The total amount of restructuring that goes on as energy moves toward entropy is therefore varied. Lotka has developed the thesis that evolution is the process by which the patterns derived from variation in the gene pools to be found in a habitat are so rearranged in numbers that they capture the maximum amount of energy. Equilibrium is reached only when this maximum has been achieved. This would provide an explanation as to why the numbers of different organisms sharing an environment varies, and also give a basis for understanding more exactly why the maximum number of a particular population in a community was fixed at the point that it was, without having to account for equilibrium by talking only about equilibrium. The evidence for Lotka's position is not yet sufficient to make clear that it should be formulated into a law. But the tendency it expresses, whether or not the process ever reaches the end point he designates, fits other evidence that ability to control energy conversion is one factor involved in the persistence of patterns that require energy for their replication. Certainly the patterns of observable human activity fall into that category. Man cannot escape thermodynamics. He is dependent upon the plants that synthesize radiant energy and inorganic materials to supply the energy in the forms upon which his own body feeds. He competes with other animals for plant life, and in the process changes the life chances for various other plants. His effectiveness in controlling the conversion of energy so that it serves his needs and satisfies his values is one measure of his probable survival in a habitat. Energy flow is, of course, not an independent variable. When we say "control over conversion" we indicate that it is being directed and the social organization through which that control is effectuated and the values that direct choice are part of the energy system. But there is no gainsaying that there is a degree of variability in the response that is given by objects in nature to the same act by humans. The amount of energy flow one gets by kindling a lump of coal is altogether different than that which results if instead of coal, quartz is added to the fire. Certain kinds of response, then, derive from certain kinds of materials. There may be myriads of social arrangements that have the effect of meeting the conditions under which materials will yield this response. If increased energy flow results from any one of these arrangements, that energy becomes a new, dynamic fact to be taken into account in understanding whether it will enhance the probability that patterns will survive unchanged or the flow will change the system that gave rise to it. Obviously such an increased flow may become a factor altering choice—it may change the hierarchy of values of those who have come into possession of it. It may by feedback reinforce all of the social arrangements that existed at the moment of flow increase except those that would throttle it. Even these arrangements that limit flow may be reinforced to a degree because the human being that carried on the arrangements leading to energy flow may alter them to stop it or slow it down, if they discover that it endangers other values more important to them. The most recent case in point is, of course, the effort to control fission and fusion explosions. What we want to point up is that the increased flow has consequences and that these consequences *can* lead to further increased flow and so on. The flow is a new dynamic fact to be taken into account. It is a means that gives direction to technology and through it those complexes immediately associated with technology, and so on more derivatively. Much of a culture may be so remotely related to changing energy flow as hardly to be disturbed by it. Efforts to direct the new flow to preserve these tertiary elements of a culture may succeed while at the same time elements primarily and secondly related thereto are forced to give way as men endowed with new power by the new energy source direct that power toward their alteration. The conditions necessary to convert increasing energy flow may in some cases (as, for example, in setting fires to burn off unwanted vegetation) impose only limited conditions on the society that uses it. In others (such as, for example, controlled nuclear fission or fusion) they may be extremely narrow and the technology and social organization required to operate them be extremely precise. Here, if the flow is to be secured and maintained, most of the arrangements are dictated by the facts discovered by science to be necessary. In a sense, then, they do not flow from values or culture but from the nature of the patterns built up during the evolution of the stars and the earth and the biosphere. The direction imposed is, to a degree, permissive. That is to say, in a sense nature says "if you want this, these are the conditions under which you can have it." A culture or a value system can cause men to turn away, refusing to meet those conditions, but that culture cannot at the same time receive the benefits in increased control over energy that it would get had it met the required conditions.

Technology is involved in, but not identical with, the energy control system. But if we examine technology in terms of the effects it has on the control of energy we can see better, perhaps, why it has moved in the direction it has taken. Technological inventions that result in bringing more energy under man's control give him an advantage in dealing with other men and with the other organisms with which he must compete for energy. Among these inventions some, because they have effects that reduce his ability through organization to survive, are selectively eliminated. What is left after a time is a set of techniques and tools that increase energy control in ways that do not otherwise reduce the probable survival of other patterns in the milieu. In the broadest sense, technological "efficiency" has no meaning apart from human evaluation. But it is possible to discuss energy conversion in terms of time and energy without making an evaluation. That is, a model can be set up showing how certain kinds of results could be secured utilizing the least possible amount of energy. The results of such a model can in turn be compared with what happens using various specific real arrangements and a statement in some such measure as percentage can be used to indicate how nearly the real approaches the "ideal" system. Also given the object to be secured and the converters to be used, it is possible using an ideal model (such as a Carnot engine) to describe the rate at which it can be secured with "least effort."

At times there have been attempts to show that this is the principle behind cultural evolution. The principle has some application at certain points but it is too limited. As we have indicated, it is only when the result to be achieved is known, the converters and the energy source specified, and the rate established that the least amount of energy required can be determined. But it gives us no basis for learning what the end results are to be, and what, in terms of all the results to be secured (or the effects evaluated) is the system of least cost. It is this that we are trying to learn.

The widest framework we could use to discover where we are headed might be Newton's Second Law of Thermodynamics. According to this law, all the patterns whose survival we are interested in predicting will in the long run disappear and entropy will take their place. This answers none of our questions. Does the law work in more limited time? That is, does energy always move in the same direction, or is the law a statistical statement such that great fluctuations in direction during shorter periods are covered up by general trends? During the time the patterns in which we are interested seem to have existed, they appear more nearly to follow Schrodinger's principle of movement from order to order than in the direction of entropy. But this generalization, too, is statistical and gives no direction for short periods or specific patterns.

If we confine ourselves to the period of his existence as such, it appears that man has been engaged in producing patterns that persist far longer and involve the structuring of energy in greater amounts than was required to maintain him and the plants that originally captured for him the radiant energy from the sun. To the extent that such structures appear to be unnecessary for this biological survival (such as the pyramids for example), man appears not always to have acted on the principle of least effort. So it is not necessarily true that he acts so as to conserve energy. Certainly the fact that his control over energy is always limited forces him to choose among possible uses known to him. But regardless of his purposes he must assign a minimal amount of energy to secure the energy with which he acts on his environment. He must also use some energy to protect his energy flow from other plants and animals seeking to use it for their own preservation. Beyond this he has choice as to what he will do with the remaining energy that is under his control, if any. Its quantity and the form it takes will have much to do with what else he can accomplish. In a given habitat the kind of tools he commands has much to do with the size of this. If he possesses only a digging stick with which to dig up roots, rodents, beetles and such like, he may have to spend most of his time and energy in this pursuit. The very limited flow of energy available will limit the complexity of the organization he can develop. Thus only limited forms of the family can be sustained in these systems.

Perhaps it is not happenstance that so very large a part of the artifacts coming to us from early man consists of weapons. With such tools it was possible to kill other larger animals and so gain control of more energy with less expenditure of his own. The development of the bow made it possible to store energy that upon its release would drive the arrow at a much higher velocity than he could attain by running. So with bow and arrow he would come into possession of the energy already captured by animals much more fleet than he, such as the deer. With the spear he was able to obtain so great a store of energy as was represented by great animals like the bison.

The analysis of the kinds of social organization that can most effectively be used to exploit energy sources of this kind has been undergoing very careful scrutiny for some time. It has become apparent that there is some degree of likeness among men in cultures using approximately equal amounts of energy from similar sources. But just how completely it will be possible to predict social patterns this way is not yet clear. Certainly the forms of social organization vary in their effectiveness, not only with the tools or weapons used but also with the plants and animals found in a given habitat. Organization and technology are inextricably combined in a complex that results in a given energy flow. A number of combinations may from this point of view have the same result. Feedback in energy terms would not distinguish between them. A broad spectrum of kinds of social organization compatible with given technology may exist. But there do seem to be "jumps" in the classifications. Certain kinds of organization require for their perpetuation minimal amounts of energy. If this is not forthcoming, the whole organization disappears. It is easy to see, to use a biological analogy, that many mice might live where no rabbit could, though they fed upon the same source. The appearance of many kinds of elaborate social organization waited upon technology that would permit control over large amounts of energy flow. If subsequently energy flow should fall below a critical point the whole system must fall apart. This thesis is not adequately demonstrated nor have the quanta of energy sufficient to maintain specific kinds of organization been worked out. There is, however, considerable historic evidence for it. Periods during which man suddenly began to build organizations far more extensive than those he previously was able to maintain have occurred. Their incidence may be called periods of "revolution," the kinds of systems that emerged may be called "stages," or given some other name. The point is that we can show how new developments that brought sudden enlargements in man's control over energy were accompanied by what appear to be almost discontinuous change. Some cultural evolutionists give the name also to other periods when "jumps" in energy are not now apparent. In any case, there are times during which a great many of the previously regularly recurring patterns that served to show cause and effect seemed suddenly to disappear or be very greatly changed. Cultural anticipations of what should happen no longer received their accustomed feedback. Organizations previously insignificant in power and function came into control of great accretions of energy and with it the means to create choices that enhanced their own survival chances. Many others that had long been dominant began to lose power, controlled less and less energy, and became incapable of carrying on even the functions they formerly performed.

There is growing agreement that control over fire, development of weapons, and the cultivation of plants and domestication of animals represented critical points in the development of culture and social organization. For other anthropologists and historians perhaps it was the invention of the development of social organization itself that constituted means to use energy in ways so much more effective as to be classed as a new source. Childe, for example, looks upon the city in that light.

Elsewhere I have developed at some length my own conception as to what were critical points in man's use of energy and the way those who passed these points were able to expand the area over which they were dominant, reinforce or change the values that affected the choices they made, control to a degree emerging social forms, and enlarge further the flow of energy by developing science and technology. It would be a work of supererogation to repeat these findings here. But some conclusions about the way they relate to evolution and to the underdeveloped areas might profitably be pointed out.

I. First we must note that when man captured and put to use energy from sources other than that secured through photosynthesis, he introduced a factor that affected vitally the previously existing thermodynamic balance. Sometimes the quantities of energy he controlled, such as those from the use of fire for example, were greatly in excess of anything that any other organism could bring to bear in its efforts to survive in competition with man. Sometimes man, by transporting them from great distances, through the use of the wind or fossil fuel, brought plants and animals that otherwise probably never would have reached there into competition with those previously found in a habitat, completely disrupting any former tendency toward equilibrium.

The result was, of course, to create situations that could not have been predicted on the basis of the balance that existed when local patterns of organization evolved. The new energy flow provided feedback from sources so far removed from the immediate situation that no amount of adjustment between things living there could restore the previous situation. Whether we examine the flora and fauna so disturbed or the *human culture whose evolution was based on feedback from it* we find that it would be completely impossible to restore the functional relationships that once worked automatically. Where the flow of energy in a habitat is greatly enlarged, culture thus becomes more and more contrived, that is, its continuity depends increasingly on man's ability consciously to understand how it works and to be effectively able to manipulate the elements that make it work. As a consequence, discontinuity in culture becomes more common because much of what was there was never understood, verbalized, or integrated by conscious process. Under the new situation, however, "letting nature take its course" is increasingly unlikely to produce expected results. With increasing frequency *values* based on man's experience during the time he was building culture based on energy from plants and animals, induce choices that in the new situation are self-defeating, and *the more energy devoted to their preservation the less likely a workable feedback system becomes.*

II. As a derivative of this the previous relationship between dominance over an area and the growth of population becomes inverted. In societies dependent primarily upon the muscles of men for mechanical energy, the only way greatly to increase that energy is to increase the number of men. But when converters using wind or water, or fossil fuels, but requiring for their own reproduction a good deal of energy, begin to be used, it often becomes true that the flow of energy can be increased more rapidly by slowing down or stopping the increase of men than by permitting or encouraging it. The energy that might have been used to beget children and support them to maturity can then be devoted to the production of converters with a much greater capacity. The

greater the disparity in the cost of producing humans and that of producing other converters of equal energy capacity, the more marked the
advantage, in energy terms, of limiting the growth of population. Once
a stock of high efficiency high energy converters exist, together with a
reliable source of fuel, the rate of increase of high energy converters can
be geometric if only the energy used to make such converters can be
directed into the production of still more converters. Because the
amount of energy over and above that expended to secure it (what I call
surplus energy) which can be secured from coal or petroleum is so
enormously greater than that which can be secured from producing food
or feed, the exponential can be, comparatively speaking, very much
larger for machines than for men.

III. A small population deriving its food from a relatively small
amount of fertile land can, with the aid of high energy converters and
fuel, become dominant over a much larger population forced to secure
its food from land of declining fertility and to utilize in production
many men who are in terms of input-output ratio much less efficient
than the machines that might replace them. So cultures that carry values
which result in enlargement of population are by those values greatly
handicapped in competition with high energy cultures not so burdened.

IV. Cultures that encourage the preservation of social *organization* that
limits the multiplication of high energy converters in order to preserve
social organization sanctioned by low energy culture reduce the probability that they can survive in competition with those that encourage the
invention of organization able more effectively to serve high energy
technology.

V. In making a transition from low to high energy technology, the
compatibility between already existing culture and that which must be
created to operate the new technology becomes a factor. It may be that
a system that depends on slow changes, undertaken only with the
consent of those who must change, will prove to be more effective than
does a more rapid one. On the other hand, experience with culture
shock, and recovery from it, may indicate that here too there is a kind
of quantum phenomena involved. A sudden shift to a new technology
that is immediately able to deliver large increases in the flow of energy
may be more quickly and certainly established on a feedback basis than
does one initially psychologically more acceptable, but deprived by its
very slowness from increasing the per capita flow of energy. The evidence is not conclusive, but it certainly cannot be said unequivocally to
support the idea that democratic change will always prove to be superior
to other types that do not require consent of those who will be involved
in it, before change is made.

If these theses stand up, we have shown how technology is given a direction by evolutionary forces, somewhat independent of the values of particular sets of men. If all men would cease the effort to come into control of more energy, they might, by the future consequences of human and other kinds of organic reproduction, be returned to the condition in which they previously existed. But if one set of men, tired of the effort to adjust to culture changes induced by increasing energy, or satiated by its product, cease to increase their energy flow, there are almost certain to be others, who have never had control over enough energy to provide the satisfactions and the power they would like to have, who will take up the process of expansion of energy flow. It will only be when we get another response from nature, in the form of greatly diminished return in the form of surplus energy, that we can expect the present revolution markedly to slow down.

There is knowledge that leads to the conclusion that the time this will take place is not so far away as we would like to think. The results of one set of investigations led to the conclusion that if the world's population were to continue to increase at its present rate, and all of it used energy at the rate it is presently being used in the United States, all of the energy available at up to four times its present cost would be gone in a hundred years. The number of such estimates and the size is variable enough that nobody should take any specific one too seriously, but the point is that energy sources, except for solar energy, are quite finite and there will be, in a comparatively short time, an end to the technological imperative provided by the response nature gives to our search for increased energy. We are not talking about time periods similar to those that produced major changes in man's gene pool. Even if only a few more of the world's people increase, their use of energy at the rate we have, easily available energy, energy yielding very large surplus, will be gone in far less time than the duration of the Roman Empire. And then what? At the moment we are building up the size of human population at a terrific rate. Part of this has come from the elimination of diseases in places where in the past the food chain was adequate to support a population limited by those diseases. When population limitation becomes generally to depend on limited food resources, it is probable that new plants will be introduced that have higher yields than those now used. But this merely shifts the limit back to that imposed by the local nitrogen cycle. Use of great quantities of fossil fuels to activate nitrogen means more rapid exhaustion of these fuels, plus further increases in a population that cannot be sustained unless energy flow into synthetic fertilizers is also maintained.

It is possible that we will learn how to control the fusion process and get energy from sea water, but since we do not as yet know the energy costs of harnessing this energy, nor the yield, we cannot be sure that it will produce larger energy surpluses than the great bonanzas we are now exploiting. Generating equipment wears out and has to be replaced so the series representing its accumulation is convergent. Increasing amounts of energy are required just to replace existing equipment. What we can be sure is that unless some other limit is placed on the expanding system represented by human reproduction, no matter how large our stock of converters other than man eventually becomes, it will not be enough.

The social systems now in control of high energy technology are not required to diffuse them over all the earth. They can instead choose to confine its spread to areas that provide the maximum energy yield per capita, given various combinations of access to scarce natural resources, leaving large areas of the world in low energy cultures. It is possible that of two social systems, that which took the latter course might survive after the first had thus dissipated the energy necessary to maintain elaborate organization and regressed to a simpler form capable of supporting only limited technology.

At the moment, acting upon the basis of values derived from an earlier day or from pre-scientific dogma and with a naive conception of the relation of values to energy, the mood seems to be that energy shall be scattered broadcast. This is done in the effort to preserve traditional ideology, to court popularity, or to preserve the image of a benevolent and well-meaning people. Whatever the basis for such a choice, it will have consequences, and in the light of these experiences, after we have had further reaction by others to our assertions about the "Truths We Hold To Be Self Evident," we are likely to make an agonizing reappraisal of them.

Other peoples are not driven by any evolutionary principle to recapitulate the sequence of events that took place here. Nor need they use all of the ideas that we used to justify it if they did not guide our success. Some of those ideas undoubtedly did contribute to it. Others, had they alone been followed, might well have reduced rather than increased feedback from energy sources. Because we simultaneously used them, all of the ideas have been strengthened by the increased feedback that did take place. But they are not immutable evolutionary principles.

If we want to know more about the functional connections that exist between our culture and social organization and emergent high-energy technology, we will have to study them far more carefully than we have done so far. Our recent experiments in trying to introduce that organization and technology and ideology in faraway places were not carefully scrutinized, using scientific methods; so we have lost a great opportunity to discover a lot about which elements of the complex that grew up here were patterns that contributed to increase in the efficient use of energy to maintain patterns necessary for its generation, and which practices only contribute to the preservation of traditional ways that are themselves parasitic in energy terms. But it should be clear by this time that neither Smith nor Marx knew much about what he was talking insofar as it would provide a guide to emergent social organization. Capitalism's fight with Communism may be over ideology, but it will be conducted with instruments and power that were produced more often on either side by disregarding those ideologies than by conforming to what they would require in the way of organization. Our claim on the minds of men may well rest more on what happens to their bodies in terms of nourishment, disease prevention and provision against the vagaries of a hostile environment than it does on the proclaiming of truths and the shouting of slogans.

This is not to say that whatever means will increase energy flow are thereby justified in terms of cultural survival. We have already entered a caveat against the materialistic dialectic. Throughout this discussion we have asserted that values, organization, technology, and the inorganic world are together involved in producing and maintaining the patterns that govern energy flow. So in a sense it is tautological to say that patterns survive because they maintain energy flow. It would be just as meaningful to say that there is energy flow because there are patterns. Neither technological determinism nor "energetic determinism" has been demonstrated. But neither has the transcendental view of values or culture. If man makes himself, he does so in a world that includes other organisms and things that interact with man's purposes as they are reflected in his choices and resultant behavior.

There are certainly limits on the kind of organization that the kind of animal man has become imposes on evolving energy systems. Moreover, the dynamic functions of the organism are involved in the responses given to stimuli. The human infant responds not only to food and sex but also to social interaction with adults. No society can survive unless it creates values that reflect these facts. What is involved in the value system necessary to operate high energy technology is only now becoming evident as we have control of sufficient energy to discover it. But if values are, as I think they are, natural, not transcendent or supernatural phenomena, then their origin and their efforts in high energy society can be discovered and heuristically it becomes increasingly desirable that we find out what they are.

It is not our task here to talk about organization, or values or organisms in evolution except as they are involved in technological progress. What we have been trying to show is that technological progress has no meaning apart from the complex which has accompanied its evolution. It will not make sense for us to offer to others our technology, designed as it is to serve our values and our organization, on the assumption that they *must* hold to be self evident the Truths we hold so to be, and *must* be seeking the freedoms we are seeking. But neither will it make sense for them to borrow our technology on the assumption that they can take it and disregard all the other patterns involved in the complex that makes our technology work. Both we and they need to know what is necessary and what is adventitious in the relationships of technology to society.

Perhaps if we abandon our own dialectic with its all or none assertions, we can find them out.

Chapter II

Technology and Social Change on American Railroads

Some theorists hold that man is inexorably driven by technology through specific kinds of social change to predictable forms of social organization. The orthodox Marxian, among others, holds this point of view. But, on both sides of the Iron Curtain, those who have tried to put knowledge about material things into operation have found no sure guide showing how to get innovations adopted with only the expected results.

The fact that technological change is always accompanied by social change is not in doubt. What is questionable is the nature of the relationship between them. Which is cause and which effect? What is the order of their appearance; must the social come before, simultaneously with, or after the technological? We cannot rely upon time sequence alone to reveal what is causal. Much of what a man does is done in anticipation of its consequences. In that case, which is causal, the thing supposed to produce the result, or the thoughts of the man who anticipated it? For this and still other reasons scientists have modified older concepts of causation. Instead the search is for propositions that will increase the accuracy of statements about the probability that, under certain conditions, prescribed events can be reliably expected to recur. The formulation of these propositions does not dispose of the necessity to observe facts and their order. It does raise a question as to the significance of sequence. All scientific law requires us to show that "under these conditions, when this occurred it was followed by this." In this sense history is a necessity. But in a laboratory the order in which the elements making up the required configuration are brought together can be altered and the discovery made as to whether order is significant. Often it can be shown that what are called "predisposing causes," to be set off by the introduction of "precipitating causes,"[1] can themselves produce the expected result if they are introduced into a situation where the other necessary factors already exist. It then becomes clear that in some cases what is required is not specific order but only the completion of a given configuration, in whatever order. It then becomes of little significance that a particular order was observed at the point where the law was first formulated.

Sometimes the same conditions hold in the development of social science. But many of the kinds of things we are trying to relate have happened in only a few places and not very many times. In these cases order may seem to be invariate. Moreover, since the bulk of social phenomena is carried from one generation to another as a single "bundle" of culture, it is hard to discover what is in a causal sense necessary for the persistence of a specific complex and what is only adventitious. As cultures are put under strain or become disorganized over time, we may find that often what seemed manifestly a causal connection to the actors involved at the time has later been shown to be dispensable. The historian reports what he sees, that all these things are going on, in this order. However, what the people involved thought necessary may not be important and they might not have even recorded the most significant factors later shown to have been inducing the behavior reported by the historian.

Until recently advanced technology was the possession of a limited number of Western peoples who had common origins, fairly similar institutional arrangements, and hierarchies of value. Since the tech-

---

[1]This dichotomy has been widely used, particularly in the behavioral studies, since it was first enunciated by Edward Glover, *War, Sadism and Pacifism* (London: Allen & Unwin, 1933). Many economists use another dichotomy in which "all other things" are supposed to remain equal while there is variation in the factor whose causal influence they are examining. In both cases an assumption is made that one set of "causes" *can* vary independently of the others. This is a prejudgment that is frequently not demonstrable either theoretically or empirically and the assumption may introduce an initial error which invalidates all subsequent manipulation of the evidence.

nology had not been put into operation elsewhere it was natural for Westerners to assume that all the arrangements characteristic of these cultures were equally necessary to the achievement and perpetuation of scientific and technological advance. As more evidence came in, it became apparent that in the long past, where a very wide range of social systems experimented with a wide range of geographic and technical conditions, change did *not* always take place in a given order, a particular kind of technological change was not always associated with the same kind of change, nor were all of the particular institutional arrangements surrounding use of a particular technology the same. Stage theory broke down and for a time any kind of secular trend or ordered appearance of social change was denied. All kinds of evolutionary models were under attack.

But the practical necessity of doing something about technological change remains. Science and technology *are* dynamic and *do* have effects. Those who would use them have to develop some kind of idea about how they should proceed. They need to avoid the massive errors that might prevent success in the foreseeable future. They need to know what the social consequences of the technology they seek to use are to be, for inevitably the effects that technology has will also affect whether it will continue to advance and spread or be cut off. Those who hold values likely to be destroyed or downgraded in the hierarchy of a society as a consequence of technological change are as vitally affected as are those who hope to achieve more of their own values and upgrade them in the hierarchy by the same changes.

For example, small farmers, businessmen, and others with a stake in rural living make strenuous efforts to prevent the advance of technology which threatens their way of life. Some, like the Amish, attempt directly to prevent the use of the new technique. Others seek to impose indirect controls, as in limiting the size of the planting in tobacco, that have the effect of rendering machine technology unusable. In other cases the effort takes the form of using taxation to saddle the innovators with at least part of the losses suffered by those disadvantaged by the change. There are also many other types of situations where "conscientious sabotage," to use Veblen's phrase, takes place. These include "featherbedding" by unions and similar but not so widely recognized actions by management. Local government itself is retained in forms not suitable for the effective use of modern technology in areas like the maintenance of health, sanitation, fire and police protection. Very often the people who resist change are unable to verbalize in ideological terms the rationalization for what they are doing. What is important for the theory of change is the fact that they *act* on the basis of their feelings about the specific issues before them, often without knowing how these acts will affect some other conditions to which they will then have to react. It would be foolish for the theorist to expect that mere demonstration that something is "technologically superior" will automatically win the enthusiastic support or even the compliance of such people.

Nevertheless, the question as to just what *must* be done if the benefits possible only with the use of new technology are to be gained in a particular society becomes increasingly important at the same time that it is increasingly difficult to discover. If the people in the older industrialized states act on a melange of ideology, experience, rationality, and anticipation, those in unindustrialized areas must on the other hand try to find some theory on which to stake their own efforts. In other words, they must fight the battle along theoretical and/or ideological lines if they are not to depend upon blind trial and error in a situation where too many mistakes may prove fatal to their cause. So Westerners who would extend aid to them must be as concerned to know what will achieve the desired technological breakthrough as are the Marxists who insist they already have the answer. Under these circumstances, "stage" theory is again invoked, and evolution, particularly cultural evolution, is once more restored to intellectual respectability. We now know that recent efforts directly to reproduce in "underdeveloped" areas certain kinds of social organization that are already stable and regenerative in older industrial states have had results quite different from those anticipated. Whether or not specific sequence *is* required for innovation, it is clear that it is necessary for certain social and technological achievements to have been made before a viable industrial system can *survive* without outside support. In the absence of these technological factors tremendous ideological efforts to produce "advanced" systems turn out to be abortive.

Yet to determine just what the required technological conditions are for the achievement of given social values is not itself easy. There is great difficulty in classifying or typifying technologies without using criteria that are at least in part social phenomena. Without some means to categorize them which is independent of the social matrix within which they operate, there is constant danger of elliptical reasoning that gives no means to assess separately the influence of the ongoing technological, physical, or biological processes, as distinguished from that of social processes. We thus cannot say whether a given system succeeded primarily because it was capable of handling technological factors effectively or because the social arrangements met the expectations of the people who used them.

A Model Using Energy as a
Factor in Change

It was to help break out of this cycle that I attempted in *Energy and Society* to develop a model in which technology could be analyzed in terms of the energy conversion involved in its operation. The thought was that if we could show by measuring energy that *alterations in its flow* on a social system were correlated with *social changes* in that system, we might thus establish a means to know something about how dependent *were the appearance and survival of particular kinds of social phenomena on the material conditions necessary to secure and direct that flow.* It would be redundant to develop the whole case made there. Since the model used there will also be used now, I must outline it.

First of all, this is a feedback model. It has recently become popular to treat almost any kind of interaction as feedback. There is some justification for this position. But the model we are using is one which depends on instruments to trace and measure *energy flow* directly in *physical terms.*

In feedback systems a part, often a tiny part, of the energy being converted by a system is, through instruments, directed to "sense" some "outside" condition, and, in terms of its findings, redirect the flow being converted through the whole system. From energy derived from the larger source, what once was a tiny influence thus comes to be a major one. So one element involved in creating the ecological balance between man and his environment is the constant appearance of variables which, because they do or do not result in increased energy flow, do or do not have increased chance for survival in competition with other pre-existing variables.

Man's control over energy is always limited. The social arrangements that maximize energy flow from the physical and biological world *are by reason of that fact* provided with one advantage in survival vis à vis others that achieve smaller flows. This is not to say that all systems found at any point will be those that maximize the energy flow possible there. Social systems that achieve higher energy flows may at the same time use them less effectively to meet required conditions than do others that can convert only smaller flows. Before the probability that any system would survive in competition with another could be accurately stated, it would be necessary to examine all of the relevant conditions. But this does not deny the significance of energy flow per se. It means that to overcome the handicap of a lower energy flow a system must have some other offsetting advantage. Obviously there will be many situations wherein such offsetting factors do exist and energy differential is not sufficient to be overriding.

But the obverse is also true, and when the energy differential is very great it may lead to the survival of systems that are in almost everything else less conducive to survival than those they replace. That is, they may disrupt the functions of the family and community, undermine the authority of the religious, result in the destruction of old economic arrangements, and destroy the ability of government long in power to rule legitimately, yet survive simply because the energy flow is sufficient to overcome these handicaps and simultaneously to provide through feedback the means to establish new institutions that better serve technology and sustain increased flow of energy.

The new arrangements in turn may subsequently be subject to competition with still other innovations that further increase energy flow or more effectively utilize it to achieve the values of those who control it. That which moves in the direction of the maximum flow a particular technology permits survives more often than does any other set of relationships which innovators have offered. For this reason the tendency to identify technological change with "progress" may be to a certain extent valid, if one means by progress, change sustained in a particular direction.

However, we must also recognize that many of the decisions made, which have this effect, were not consciously directed to this end. Energy flow is not per se the objective of many men. But what they do seek is usually to be obtained only with the expenditure of both time and energy. Increased energy may be used either to secure more things in the same time or to reduce the time which must be spent to secure things. The effects are not uniform. New technology in one field may permit the use of such an increase in energy that costs fall tremendously, while at the same time to do something else takes just as much time and energy from old sources as previously. If people contemplating the changed costs either decide to get the old ends by new means, or to choose as ends more of what has become less costly and less of what is now relatively more costly, they are altering their value hierarchies. They proceed to provide social sanction for the new roles necessary to make the new technology operative. Moreover, in providing new sanctions they create a new morality and value hierarchy that is then passed on to succeeding generations. What was originally to be secured through individual experience in adjusting to new technology is now obtained through cultural transmission.

Heuristically it is justifiable to study separately values, ecology, technology, and culture but those who study the behavior of men in changing situations will always find change going on in all of the categories of evidence they study.

Analysis of social structure or the institutionalized means to use particular kinds of technology will show that here, too, the existing arrangements are being reinforced by feedback from the system. As innovators produce new variables in order to use a particular kind of technology these variables are either reinforced by feedback because they enlarge the flow of energy or are put in jeopardy by their failures to do so effectively. For institutions that do permit increases in energy flow, simultaneously put into the hands of those who control them increased means with which to seduce, corrupt, or coerce those who would oppose them.

When we say that this is a feedback model we must of course recognize that it is not a complete explanation of social change. We do not go into the nature of the relation between the origins of specific innovation and its selection for survival or elimination, though obviously, as we have already pointed out, there is some relationship between what is anticipated to be the consequence of an act and the choice either to perform it or not. Nor does the theory attempt to deal with change not dependent on or responsive to changes in energy flow. The emphasis here is rather on the "law of effect." That is, we are concerned with classifying actions in terms of what happens *as a consequence* of man's having done some particular thing.

It is clear that some of what happens could only have happened because men reflected on what they had observed, speculated about it, and decided on the basis of their own values whether to repeat past acts or undertake new ones. Whether they are able to act in a particular way and still be within the limits permissible in their society, with social approval, or must violate norms to put new inventions into action will, of course, be part of the anticipated effects. It is not only technological gains or losses that affect the innovator's decision, nor are the effects of his acts confined to the realm within which the innovation is made. So, for example, an innovation which may in technological terms be fully justified may nevertheless be exorcised because of its putative social or religious consequences.

Obviously the historian who surveys what has occurred during a particular period may select any of a number of changes as being significant. He may choose a starting point and establish an order during which changes took place and indicate a preference as to which is "causal" in that sequence. It would be foolish for us to quarrel with those who cite an institutional invention, the emergence of a particular power structure, a new value hierarchy, or the juxtaposition of a man and a situation as the base from which to explain a particular change.[2] But if common factors can be shown running through some of these circumstances, these factors may provide a clue to understanding not otherwise provided. For this reason empirical studies of a large number of situations which have in common that they involved a particular kind of technological change may, when made, demonstrate the usefulness of studying the flow of energy as one important variable.

Railroads Provide a Historical Case Study

Analysis of changing technology and other change on and around the railroads of the United States should provide us with one example. Railroads were among the first organizations drastically altered by the introduction of fossil fuel. Because they have a longer history than most industrial organizations they may show more about longer-run relationships than do newer ones. The fact that it is transportation rather than, say, manufacturing, that is involved may render a good deal of what happens irrelevant.

The history of railroading occupies an enormous literature whose presentation and analysis would require too much time for the insights it provides. To make it meaningful without resorting to this heroic process we must find some things common to all the railroads and see what the recurring relationships between them are.

Let us now utilize the general theory outlined above to see if it aids in this endeavor. Let us see what changes in the flow of energy and in its source may have to do with the introduction and spread of the railroad.

---

[2]Adolph Berle and G. C. Means showed how the development of the corporation was instrumental in providing a framework which made the effective use of modern technology possible. (*The Modern Corporation and Private Property* [New York: Macmillan, 1933]). Max Weber is an outstanding proponent of the proposition that the value system of a society has a great deal to do with technological and other forms of change. (*The Protestant Ethic and the Spirit of Capitalism* [New York: Charles Scribner's Sons, 1930]). See also particularily the work of Pitirim Sorokin, Howard Becker, Talcott Parsons, and their students. Among those who emphasize power as a factor Bertrand Russell is a modern pioneer. (*Power, A New Social Analysis* [New York: W. W. Norton, 1938]). Others prominent in this type of analysis include Felix Gross (*The Seizure of Political Power in a Century of Revolution* [New York: Philosophical Library, 1958]), Franz Newmann (*The Democratic and the Authoritarian State* [Glencoe, Ill: The Free Press, 1957]), and Hans Morgenthau, Barrington Moore, Jr., Richard Schermerhorn, Harold Lasswell, C. Wright Mills and many others. The "great man" theory of change is so ubiquitous that citation is redundant. As it relates to technology, biographers of men like Henry Ford, Andrew Carnegie, Thomas A. Edison, and their like would start their interpretation of change with the appearance of these innovators.

The first railroads were *rail roads.* That is to say, they were roads on which, instead of paving the whole surface, a road bed was built with sleepers or cross ties supporting a rail, originally of wood or stone, later of cast or malleable iron, finally of steel. For quite a while the cross ties were buried deeply so the draft horses used as prime movers could find solid footing, without damage to the ties. The gains came in reducing friction between the rolling wheel and the surface it traveled upon. This permitted the horse, and later the locomotive, to move larger loads, at higher speeds. At first the means to produce tractive effort, the horse, remained the same as on other roads. Where a road was heavily traveled, decline in operating costs was great enough to more than compensate for the greater cost of roadbed and rails, but many roads, built to fit local needs, were more economical to use than was the more costly railway.

In fact, as one looks at the shift from road to railroad it is hard to see the advantages it offered. The engine was less efficient in energy terms than was the horse; the horse was able to deliver at the drawbar about 20 per cent of the heat value of the feed it consumed, while the steam engine was not often, in the same terms of fuel input and mechanical energy output more than one-half of one per cent efficient. The engines put more weight upon the rails, requiring that they, and the roadbed, trestles, and bridges supporting them, be more strongly built, at higher energy cost. The cost in terms either of the human time or the energy consumed in building an engine was considerably greater than that involved in raising a horse of equal tractive power, and the life of the engine was often less than the working life of a horse. The horse was far more versatile than the locomotive and its use could be shifted from railway to road, or into the field, as occasion might demand. And initially, and possibly even today, the cost of maintaining the kind of social organization required to use the locomotive effectively was greater than that involved in the more simple social structure adequate to exploit the horse fully as a prime mover. It is not surprising that there were many who could not see the advantage the locomotive offered. In terms of the economics of the day, little seemed to be gained by adopting it.

If the theory outlined above is correct, the solution is to be found in the difference between the sources of the energy used by the two techniques of propulsion. The power exerted by the horse originates in the radiant energy of the sun. Through photosynthesis in plants, this energy is made available to the animal's muscles, which convert it into mechanical power or work. This process is relatively inefficient. The best plants (in these terms) convert only a very small part of the radiant energy of the sun falling upon the land they occupy into forms of plant life that can be digested by men and plant-eating animals. Even a good system of cultivation on fertile land seldom yields more annually than ten times the energy required to sustain the cultivator and the offspring necessary to replace him. Very frequently the yield is just enough to sustain those who gather it.

The difference between the energy input involved in securing energy and the resultant energy output we call surplus energy. The excess is, in energy terms, free. We have gotten from nature more than we put in. There are very few kinds of human activity that have this result. Most of the time we put energy in, and have less than we formerly had. We do this because we value more highly the things we achieve or secure than we do the energy we expend in getting them. Thus a man might spend a great deal of energy to get a diamond weighing an ounce, far more in fact than he would to get coal weighing a ton. In value terms the surplus gained from the work spent securing the diamond is much greater than the surplus gained from the coal. But in energy terms the coal might represent thirty thousand times as much as the diamond. We must distinguish clearly between the two kinds of surplus. It is often because of failure to see the difference that men make what seem to be economically sound judgments which are in energy or survival terms disastrous. What Stephenson and the other pioneer builders of locomotives were up against was the difficulty in explaining why, in spite of the fact that it seemed uneconomical to those using the culturally approved system of thinking, the locomotive would replace the horse. Since costs and gains are measured in terms of value, the technological base for the productive system was often not clear to those using it. They measured inputs in terms of price and outputs the same way, and if they thought about it at all, generally assumed that the "just price" which had traditionally prevailed was something independent of or prior to technology and would always show man how to apportion the factors of production most efficiently.

What the steam engine did was to make it possible for man to get mechanical energy from sources hitherto not thought to be available. It could be used to turn the energy of plants like trees, which neither man nor horse could eat, into work. While in the long run this source was limited by the annual growth of trees, it was possible at first to cut them at a rate far in excess of replacement and so tap a new enlarged energy source. But far more importantly it made possible the use of coal to do work. Even when coal first came into use a miner could mine equivalent to a thousand or more times the energy he put into securing it. And with increasing efficiency in mining operations and the increasing tractive effort the locomotive was able to exert per ton of coal going into the firebox, the rate of energy return became, in comparison with that secured from plants, fantastic.

Even the driver of a ten-horse team, each producing less than a "horse-power" as measured in the terms used to establish the capacity of the steam engine, had to use a relatively large amount of human time in the care of the horses and while driving—as compared with the human time spent on the job by the engine driver. All the energy directed by the man holding the reins came from plants with their low yield of free surplus energy. On the other hand the man at the throttle controlled ever increasing amounts of free energy derived from coal, and even with the low efficiency of his engine, was able to secure more and more ton- or passenger-miles per working hour. By using steam man could get much more done in the same time—or do the same thing in much less time than was previously possible. Thus there now existed a potential way to secure transportation at less sacrifice of the other valued goods and services, since some of them could now be secured with the time freed from providing transportation.[3]

There was a great reward awaiting those who would change the existing social, economic, technological, or political arrangements that stood in the way of this achievement, and a corresponding penalty to be levied against those who resisted such change. The reward was a direct result of the increase in free energy. It need not be a result of heightened effort, superior motivation, or, except as such knowledge was instrumental to the achievement of the new technology, new knowledge. To that extent, then, we have an "outside" factor with which to account for much of what took place as railroading worked its miracles on the face of the earth. Note that this reward was available to whomever could make the necessary technological arrangements, and the social means to permit them to work, regardless of other aspects of their social system.

In a review of the railroad era we might look at evidence of the way the changes taking place were explained, we might talk about the changing value system that accompanied the change, we might look at the social structure growing up, in relation to the substitution of coal and wood for the energy sources previously used. And we need to discuss the specific social processes involved in making those changes.

---

[3]An enlightening story of the early development of British railroads, built around the "great man" theory, but not confined to it, is L. T. C. Rolt (*The Railway Revolution, George and Robert Stephenson* [New York: St. Martin's Press, 1960]). For an up-to-date view of what is now happening on British railroads see Geoffrey Freeman Allen (*British Railways Today and Tomorrow* [London: Ian Allen, 1960]).

Original Institutional
Framework of the Railroads

The first American railroads were replicas of those found in England. It was there that the technology had its roots. Similarly, it was there that the first conceptions as to what were the desirable and necessary social and political arrangements for the operation of railroads originated. At the time the railroads were abuilding many aspects of American life were changing rapidly and theory borrowed from the old countries soon had to be altered to explain emergent social forms. What must not be forgotten, however, is that many of the same kinds of changes were being wrought in Britain and on the Continent. There, too, the struggle to control and use the rails was forcing a marked alteration in institutions. The outcome was often different than in the United States because of many factors, including particularly the ecological consequences of the ratio between land and labor and its location. It also was different because of the early military significance of railroads in European logistic arrangements. The interlacing of the Continent by railroads forced a realignment of power structures and the appearance of many new kinds of coordinating agencies that contravened the "fundamental" basis of European politics and economics. Some of these developments gave rise in the United States to new conceptions of the nature of effective and desirable organization. But for the most part it was British experience and institutions that came to characterize American railroads.

Of great importance to understanding these arrangements is the realization that the British had, over a comparatively long time, been shaping their system to encourage and support trade at a distance. The sailing ship reduced transportation costs far below those possible with any other existing means. Trade was the most effective way to use the wind, a form of free energy, to increase productivity by taking advantage of ecological differences. But to use it a whole host of changes in British culture had to be made. Perhaps greatest among them was the increase in the use of pricing to provide coordination between people at points distant from one another and sharing different cultures. They also needed new concepts making such relationships meaningful, and a set of rationalizations for their use. In an unpublished paper delivered at the Fifth World Congress of Sociology in Washington, D.C., in September 1962 Elias shows how new concepts of what was "economic" emerged. What were regarded by early British economists as being universal aspects of human nature were of course relatively recent developments confined to limited areas. Although they seemed to be unaware that this was so, what was going on in Britain had a lot to do with the models found to be acceptable to theorists there. In the meantime, however, technology was reinforcing the trading system through feedback. Pecuniary relationships replaced in large measure the economic arrangement common to manorial society or the self-contained village community. Pricing proved to be an effective way to reassess the worth of human services as men were forced from traditional roles into others made necessary by the new technology. What became dominant, at least in urban Britain and among the new elite, was the idea that it was not the new technology and increased energy flow which was responsible for the power and prosperity they enjoyed, but rather that these were due to the operation of the market, freeing men like themselves to make choices they found it profitable to make. They established to their own satisfaction that in pursuing profit they would be guided by an unseen hand to make choices that maximized the wellbeing of the people.

The common law came to represent this view. Monopoly was regarded as being dangerous and undesirable, even monopoly held by the state and operated "in the public interest." The doctrine of conspiracy provided a means to prevent the effective organization of labor, while the fiction that the corporation is a single person not dangerous to the operation of the law of supply and demand permitted the aggregation of great power in the hands of the managers of giant firms.

Many American businesses had their origin among British entrepreneurs, and for a long period of time American enterprise functioned in the framework of British law and culture. So it was not surprising that as railroading developed in the United States it operated under social forms sanctioned originally by British experience. Private ownership seemed to be the only natural way to run railroads. Railroad managers exercised the prerogatives typical in other private businesses.

The traditional logic of property required that the managers direct as large a segment of the increasing flow of goods and services as possible to the "capitalists"—those who had in fact put up the money to build the railroads. But it was not long before a new logic, which legitimatized the retention of the larger part of the new flow of wealth by the promoters, emerged. The actual investors were often paid only what was required to keep them investing in railroads rather than in competing enterprises. Since the bulk of the rest of economic activity was still dependent on energy from food, feed, and muscle power, the railroad promoted in America did not need to pay much more than could be gained from these enterprises in order to attract capital. Nor was it necessary to pay men whose other alternatives were to be found in tilling the land or working in less technologically advanced industries any large part of the product gained through their cooperation as laborers. Its abundance as compared with limited means of transportation made it possible to secure land free, or even to be paid to accept and use it.

What we are saying is that it was not necessary for a promoter to "exploit" any of the groups traditionally cooperating in enterprise in order to get rich building and running railroads. Each could be provided, in return for his services, reward greater than any he could secure elsewhere, while at the same time the shipper and the traveler could be given far better service at much lower cost than he could secure any other way. And the outcome would still leave enormous wealth in the hands of those who controlled the railroads.

This, put another way, is to say that the traditional value system in use prior to the railway age had no means for moral distribution of the wealth produced using the larger flow of energy which the new technology made possible. The old institutional arrangements often reflected an ecological pattern that could ignore energy brought from a distance. Once such energy was available, much of the justification for the old arrangements disappeared. Nor could the old system effectively decide *who* was to benefit by *how much* from the increased supply of free energy. Nevertheless, the old institutions persisted and their defense was rationalized by new arguments.

But almost as soon as the increased energy flow appeared there also appeared claimants on it and its product. Each claimant used some part of the traditional morality to justify his claim but none recognized as valid the "distortion" of tradition required to legitimatize the claim of the others. What we see if we look at developments over time is a continuing struggle among groups serving and served by the railroads —each trying to maximize what they value, at *their* least cost. The effectiveness with which they can do this is related to the strength of the structure through which they operate, their strategic location in the system, the position of others in the society, and the changing value hierarchies of the people living there. What must not be forgotten, however, is that effectiveness is also dependent upon the degree to which the decisions made by railroad management permit the exploitation of natural resources, science, and technology and so increase the flow of energy that they can use to obtain the objectives of the various contending groups.

Management performed different functions for each of the groups affected by the appearance and development of the railroads. For the traveler and shipper primary interest lay in fast, certain service at cheap rates. For the investor it was desirable to secure high dividends with little risk. The worker tried to maximize wages, fringe benefits, and security. Local and state governments came to look upon the railroads as a lucrative tax source through which they could claim for the polity its share of the wealth arising from the new technology. The military branches were more interested in the contribution a railroad could make to achieve the functions that were their primary concern. And finally, whether or not management could secure high returns for itself depended upon its ability to limit the success with which other groups achieved their goals.

Power Struggles Among Groups

If we look at the resulting arrangement we might usefully describe the railroad as the locus of a cluster of functions, or call it a multi-functional institution. Its workings depend at once upon conflict and cooperation among the groups for which it operates. Generally, in our society only the latter process is considered to be "good" and is given positive moral sanction. Conflict is generally regarded as being "bad" and it is assumed that, if possible, it should be extirpated. But since every group seeking to use the railroad to maximize its own returns comes into conflict with others who would have it perform differently, conflict can no more be exorcized than can cooperation. Instead, institutional means are developed to control conflict and limit it to acceptable forms. Social organization prescribes an arena, sets the rules, and determines the referee. At any moment these all reflect the past power of the contenders. But if, over time, there are important power changes among them they will seek to alter the arena, for it has much to do with the outcome.

It is, for example, important to know whether under the rules the vote of one man will equal that of another, or whether one dollar or a share of stock has influence equal to that of another. More commonly, some administrative arrangement is set up to weigh various kinds of consideration and determine at least tentatively what the outcome is to be. The determination as to who will select the referee to enforce the rules and perhaps declare the winner has a great deal to do with the way various groups fare. Similarly, the kind of strategy and tactics used and the effectiveness of various weapons will relate closely to the structure in which they are to be used.

The power struggle involved in the control of railroading has not often depended primarily and immediately upon the capacity of the contending groups to exercise brute force. This power has, for the most part, been relegated to government. What groups *have* done is to exert various kinds of influence in two directions: (1) to shift decision-making into an arena in which they have the maximum control; and (2) to persuade those who make decisions to make more favorable ones. In this latter endeavor they may depend upon deep-seated and widely held ideas and ideals, upon bargaining in terms of immediate and pragmatic propositions, or upon the self-interest of the decision maker and those who put him into his strategic position.

The Position of Management

In the case in point those who make decisions in the first instance are railroad managers. It is in their power to determine, at least tentatively, and within limits, the policies to be followed by the lower echelons. They will be selected by those who are legally in a position to choose them. Thus they can be expected to share the value hierarchy characteristic of the groups legitimately entitled to control. But this legitimacy is itself a function of the values widely shared in the society and the resultant sanction of the decisions made. So long and only so long as there is widespread acceptance of the legitimacy of the power they exercise can managers expect to manage effectively. As we indicated earlier, a part of the heritage from Britain included the idea that railroads "naturally" are to be treated as private property. In contrast (merely for example) the railroads in Germany were at the outset regarded as primarily public in nature for they served the strategic and political objectives of the emerging German state.

Given the idea that railroads are and should be private property, managers can rely on all the institutional means generally used to support and protect the prerogatives of the owner of such property. Like all other forms of "big business," railroad managements have benefited from the existence of this widespread belief. The identity of the corporation as a person and the use of the Fourteenth Amendment to the United States Constitution to guarantee such "persons" rights that originally were supposed to be vested only in a man with a "body to be kicked and a soul to be damned" has greatly affected what are managements' legitimate prerogatives. Similarly the development of securities with a government-guaranteed priority to earnings means that today on many railroads the prior claims of the bondholders leave little for other claimants on earnings to fight over.

Weapons of Shippers and Passengers

But if they are weak in the economic arena where law and tradition gave advantages to management and investors, other groups concerned with railroad services have developed strength in the political arena. For example, the rapid westward movement was dependent upon the use of railroads to carry farm products to the eastern seaboard and the docks from which ships could take them overseas. And as farmers and allied groups grew in numbers and in wealth they began to shape new institutions better designed to serve their own interests. That these movements brought them into conflict with the "moneyed interests" of the East and the courts that maintained the sanctity of these interests history has long since been demonstrated. Among the most significant determinants of Western prosperity were freight and passenger rates. Left in the hands of management to exercise this control as they saw fit, this power meant life or death to chosen commodities, communities, and firms. In response to what they considered to be the abuses of this power the legislatures of the Western states began to reshape the institution of property as it affected railroad operations. In spite of the drag imposed by tradition-minded judges who declared many of the newly proposed institutional arrangements sanctioned by farmer-dominated state legislatures unconstitutional, the prerogatives of railroad management were inexorably reduced, particularly as they related to the carriage of agricultural products and the serving of farmers' other interests. The Grange movement and the laws it sponsored were in part nullified by the actions of the courts, but in time the interests of other shippers induced them to enter political coalitions with the farmers at the level of the national government. The result was the creation of the Interstate Commerce Commission. Most of the shipper and passenger interest was thus entrusted to a new agency. The arena in which rates are set and schedules authorized, and the character of the services to be rendered was modified. No longer was it necessary to hold or control large blocks of stock in order to affect these policies. While management retained some initiative the veto power was found to lie elsewhere. If railroad owners and managers were to wield the kind of influence they once held they now had to overcome the political power wielded by their opponents in the United States Congress which set up the I.C.C. and the growing bureaucracy that actually performed most of its work. Short of doing that they had to face the fact that management's prerogatives were greatly reduced, and operate within these new limits.

Numerous studies have indicated how control over the I.C.C. has shifted and the consequences of these shifts. What concerns us here is that without changing the definition of property or modifying the values attached to the symbols representing it, the functions of property owners and managers were changed.

A similar movement which came to fruition later and was, at least until very recently, less complete, was set in motion by another of the groups for which railroads function, i.e., Labor. Later we will discuss in detail how this has affected management's prerogatives. At the moment we note that the social consequences of the use of power over new energy flows by those initially entrusted with its use were such as to alter greatly the future power and prerogatives of railroad managers. Nor were these social factors the only threats to be faced.

New Technology Alters Group Powers

As we indicated earlier, it was initially the rise of new technology directing greatly increased energy flows that, by feedback, strengthened social structure insuring the power of railroad managers. Further developments in technology have greatly weakened this influence. The initial development of steam power was most significant in the field of transportation. The economic gains resulted largely from the fact that with cheap transportation it was possible to take advantage of regional specialization, which in turn permitted more complete human exploitation of ecological differences. However, as the stationary engine was developed it became possible to use increased amounts of energy in local production, and sometimes the resultant reduction of costs made transportation of goods from a distance less attractive. The pattern of aggregating population in densely populated cities reduced considerably the necessity for railroad transportation relative to the volume of production. If railroad transportation cost too much the alternative was not, as in the past, to resort to muscle power, but rather to increase local use of energy from high energy sources. To serve the stationary engine some fuel was brought to the rising urban centers by rail but the electric grid, the pipeline, ship, and barge could often provide a cheap alternative. These new forms of transportation could also be substituted in other ways for the railroad. Of greatest importance was the appearance of the internal combustion engine and road traffic. Later other forms such as the airplane also provided new alternatives. This increased freedom of choice by passenger and shipper altered their power position vis à vis railroad owners and managers, railroad workers, and governments. But while the general effectiveness of the forms of control it could use was further reduced, it was still possible for railroad management to make certain decisions in response to the value hierarchies of those who had the power to select and remove them.

Nor were the effects of the new technology all of a character to reduce the choices which management could effectively make. Some of them opened up the possibility that management could, with their use, overcome the barriers imposed by the political and economic power of other groups. The diesel was borrowed from industries wherein it had reached a high state of technical efficiency, and with its use the employment of manpower was greatly reduced. Electronic data-processing offered further opportunities to escape the claims made by unions in behalf of men who now need no longer be hired. Automation in switching cars similarly provided a way to avoid the power of the unions. New forms of communication made it possible to dispense with people at places where their presence permitted local government to levy taxes designed to support the community.

What we have just been talking about are *technological changes* which served to modify the forms of social organization effectively used in railroading. We are saying that if technological factors only had been at work (and of course this was not the case), railroads would still have had to make great social alterations.

But the changes just cited did not occur overnight, nor were they independent of other factors in the railroad complex. Their continued development depended on continuous feedback from the results of their adoption. Many of them might not have occurred, or certainly would not have occurred in that order nor with such magnitude without this continuous process of interaction. Those who instituted change did so in contemplation of the opportunity costs involved in that change. They could, for example, greatly reduce costs by replacing manpower only if that manpower was expensive relative to the men and machines that replaced it. The more costly the man, the more it paid to find a replacement for him, and often the replacement was possible only if a new technological set-up could be made. Very often his cost to the railroad was as much or more dependent on the union to which he belonged, and the strategic advantage held by that union, as on any other factor. This advantage might in itself be related to many factors other than the possession of the technical skills necessary to do the work. Similarly, costs imposed by taxing authorities might in some cases make it profitable to replace men who, in order to carry on their expected functions, had to be located at a particular place, with technical equipment that did not require that the man in control be at the same place.

The list of conditions that might affect the judgment of the managers who decided whether or not to introduce technological change could be endless. The point we are trying to make is that technological invention is often created in the effort to lower costs. Some are a function of technological factors themselves, like the differences in engineering terms of efficiency between a reciprocating steam locomotive and a diesel engine, but perhaps as often the costs are socially imposed and technology is devised to escape these *sociological* barriers to the attainment of lowered cost. Moreover, we must not forget that technology is going to be put into operation only because some persons made the decision to do so. The factors affecting human judgment are seldom purely technological in nature. The resultant technology then can never be thought of as a "thing in itself," even though it may be true that over time refusal to make certain kinds of technological change will result in the disorganization or dissolution of the system rejecting it.

It is particularly important in the instant case that we see what were and are the *power of those who make the decisions* over technology, what are *their* primary interests, and what controls are exerted over them by other groups. If, for example, those who actually control the railroads have little to gain and much to risk from technological change, they are much less likely to accept or promote it than is the case of industries in which the rewards from such change are available in large part to those who introduce them. Similarly, there will be a different decision if those who might otherwise be inclined to introduce technological change are barred from doing so by others, who while they cannot themselves assume the responsibilities for management, can and do set limits to the time and place and manner in which change can be introduced, or can seize for themselves a large part of the reward for change which technology would make possible. So, again in the instant case, managers chosen by those who are, in the traditional definition of property, legitimately entitled to decide on technology, are faced with laws, administrative decisions, or union power resulting from choices made at the polls or as a result of union politics. They will in consequence make different decisions about technology than they might in the absence of control by the other groups contending for power.

As we have to a degree already indicated, railroad managers have been selected as a consequence of the evolving power structure controlling the firms for which they work. In time those who represent certain of the interest groups for whom the roads perform become more powerful than others, and as they do they maximize their return while minimizing that of other groups. Management composed of those who seek "profits" are concerned that the property be used in such a way that the "costs" of providing returns to all the other groups are kept to a minimum. They may be interested in liquidating an unprofitable service which is able to provide wages, taxes, cost of supplies, and adequate transportation to the traveling public. The other groups cannot rely on management to make judgments in *their* behalf and they have resorted to various means to maximize their own concerns. So, for example, when railroad managements repeatedly made decisions that forced their companies into bankruptcy they found that the only way they could get capital was through issuing bonds which the *government* guaranteed would have first claim on earnings. As the proportion of equity capital declined, the interest of the bondholder came to loom larger and larger in railroad directorates. Since the bondholder does not share earnings beyond his guaranteed return his representatives are less impressed with the opportunity to earn additional profits through new technology than a profit-sharing stockholder might be.

Some railroads have large holdings in real estate, some of which could bring in much greater profit if used for other purposes than railroading. Directors interested in maximizing the return to those whom they represent would logically seek to put these resources to this more profitable use. But the past abuse of railroad management has greatly limited what railroads may do. Diversification is not permitted to go very far. Thus it is more profitable for the railroads to dispose of their real estate than to preserve it in use. There is clear evidence that this is what has been and is being done.

To those who identify themselves as "the public," usually shippers and passengers, the idea that railroads should curtail service to them in the interest of anybody else is anathema. Since they do not, under the logic of property, have representation in management itself they resort instead to agencies over which they do have some control, such as the I.C.C. and the state public utilities commissions. But here their influence on technology is primarily negative. They do not have the means to induce innovation. Their activities, designed to preserve service at any cost, often deprive management of funds that might be used to innovate. The primary influence on innovation then is only to induce such changes as will reduce costs rather than create new services.

If the railroads themselves undertake innovations they have to share the profits with all those who have in the past established a claim on earnings, the bond holders, stockholders, pensioners, and tax collectors. But if a private company is set up to make innovation and sell to the railroads, the innovators can claim the profit for themselves. The result is of course that innovators have no particular interest in being represented on boards of directors, so long as the industry is healthy enough to pay their bills, and is willing to accept a rate of innovation that is most profitable. Consequently, the industry has fallen behind the rate that characterizes its competitors in transportation. This threatens the whole industry, so railroad suppliers, being no longer able to depend on the railroads as a market, must now themselves diversify or interest themselves in getting sufficient control of the roads to induce changes that would assure more successful competition. In their endeavor to make this kind of technological change they are faced with security-minded management concerned more with liquidating the unprofitable part of their business than in expanding. Some of the suppliers (like General Motors) have a large interest in the railroads as a market, as well as a service. One might anticipate that they would buy in or otherwise gain enough control to protect their interests. So far, however, they have apparently been content to improve the things they have to sell at a comparatively high profit and let existing management decide the future of the railroad industry.

Government has a great interest in the industry too. Originally it was an active partner but in more recent times it has not participated in railroading as it has in road traffic, marine shipping, barge lines, or air transport. Much of the original impetus it supplied through grants of land and other aids has for most railroads long since been dissipated. The military branch has provided and continues to provide a great direct aid to innovation in other forms of communication and transportation, but it has done little to supply research and development to the railroads. Nor does government policy offer incentive to those roads who seek to innovate. Through control over freight and passenger rates it often has the effect of penalizing efforts of those roads that try to gain new customers by providing better, faster, and cheaper service.

So management that holds the purse strings tight, enters no new ventures, eliminates marginal services, and accepts innovation not to gain profit from new service, but only to cut costs, is apt to be enthroned on all but a few roads, and so long as they serve the interests of those who select them, can be expected to operate this way. The way technology develops is affected not only by the value orientation of management but also by the way it has been structured. Like many other aspects of urban industrial society the railroads operate under very close limits in terms of time. Social structure must be so designed so as to assure that things and people are where they are supposed to be at the required moment. This is also characteristic of military action. Many American railroads were built and first operated by men who got their training as Army engineers. The military organization with its great emphasis on timing and discipline fits the requirements of railroad technology well enough that it generally has survived as the bureaucratic model for the industry. Not only are there structural evidences that this is so but the very language of the military has become a part of that familiar to railroaders. The table of organization, the relationships between line and staff, and many other parts of the system fit into a hierarchical pattern. It is difficult to innovate in such organizations. To change things means that new roles and statuses have to be created and new administrative arrangements made to sanction the behavior of the bureaucrats. The innovator is confronted on all sides by people with good reasons for "not sticking their necks out."

Not only is there resistance to change from individual railroads. The standardizing agencies like those of the Association of American Railroads have to take into account that rolling stock other than motive power is generally interchangeable over the whole of the United States. Nothing not acceptable in interchange can be sanctioned. This means that the less progressive and poorer railroads impose pretty much of a veto over the other roads. Many technological improvements must wait until these roads have earned enough to pay for them; but without these improvements the prospect for profits is bleak. The kinds of improvements that *are* adopted are likely to be those that reduce costs rather than generate new traffic. Here again social and technological considerations become merged in the minds of the decision makers.

We can now summarize the position of the various groups. Stockholders have initial legal control over management and can initiate policy. However, a large part of the control is exercised by banks, trust companies, and similar fiduciary agencies in the interest of those they represent. Control is exercised in the name of the shipper and traveler by various government agencies, and the alternatives provided by competing forms of transportation set further limits on managers. These leave them little range in which to maneuver. The result is that labor, the remaining large interest group, becomes a major element whose costs can be manipulated in the interests of those to whom management is responsible.

The Increasing Use of the Political Arena

As we have already indicated, the doctrine of conspiracy greatly handicapped the organization of unions in the United States. The individual worker was left "free" to bargain with the company. Very rapid growth of population, due both to natural increase and immigration, supplied labor in ever increasing numbers. If the native-born American, free to take to the soil or enter another enterprise, demanded too much it was always possible to turn to an immigrant who had fewer alternatives. So railroad wages need not be much—if any—higher than those in less technologically advanced industries. Efforts to create a powerful industrial union of all railroad employees were easily defeated. A few craft unions which had managed to gain power in other industries made successful efforts to organize their fellow craftsmen on the railroads, but for the most part these failed. The "operating" craftsmen, enginemen, trainmen, and switchmen, for whom there was no counterpart in other industry, could not even lean on so frail a reed. But they did manage to organize through an expedient not barred by law and the courts. They began with groups which were primarily insurance companies. The injury rate on the railroads was so high that most of the insurance agencies refused to accept employees as risks. So what amounted to "burial societies" were formed by various sets of railroad workers. This brought an organization into existence to serve their interests, and around this initial function new ones were joined as power and opportunity permitted. Finally, they reached the point that they could effectively threaten to strike. Collective bargaining based on economic considerations became possible, and theoretically, desirable. But those who felt themselves absolutely dependent on the services of the roads found such a threat intolerable. They resorted to their power in the legislature to minimize it.

In 1888 President Cleveland signed a law which provided for voluntary arbitration of railroad labor disputes. That failing, there was to be a public investigation. During the ten years' life of the law the provisions for arbitration were never used. During the Pullman strike there was an investigation but it will be recalled that this strike also led to intervention by the federal government. That the government would, if necessary, use troops showed that the public regarded the economic solution to railroad labor problems as one so endangering their interests as to be intolerable. The plain inference to be drawn was that if the unions were strong enough to cripple rail transportation effectively, government would intervene and remove the dispute from the economic to the political arena. In the aftermath of the Pullman strike, the Erdman Act was passed in 1898.[4] This act initiated the policy of having government conciliate and mediate labor disputes on the railroads. The Commissioner of Labor and the Chairman of the Interstate Commerce Commission were required to make themselves available for these purposes when called upon to do so by either management or labor. Failing to arrive at a solution through such means, the government was expected to provide a board of arbitration. Since only the operating unions could in fact effectively strike, the Act was made applicable only to them. It failed to satisfy any of the groups sponsoring it and was re-

---

[4]The Erdman Act was approved June 1, 1898. It was signed by President McKinley. It is Public Law No. 115 entitled: An Act Concerning Carriers Engaged in Interstate Commerce and their Employees. It went out of effect when replaced by the Newlands Act which was approved July 15, 1913, signed by President Wilson. It is Public Law No. 6 (§2517) entitled: An Act providing for Mediation, Conciliation and Arbitration in Controversies between Certain Employers and their Employees.

placed when the Democrats came to power by the Newlands Act of 1913.

This law created a permanent Board of Mediation and Conciliation and some full-time agents to act for it. To avoid the "one man rule" which occurred when two partisan and one "neutral" board members made decisions, the Boards of Arbitration were increased to include six members. A continuing interpretation of collective agreements was thus assured. During its life this Board was given rather wide use.

We note that between 1898 and 1916 a new arena for decision-making concerning railroad labor was emerging. Without any formal or ideological redefinition of its prerogatives, management was required to deal collectively with its employees where they belonged to unions and sought collective agreements, and to submit some of its administrative decisions to final judgment in terms other than those of success in the marketplace. The necessity to keep the technology operating had proved to be a more potent force than that of acting in conformity with ideological propositions about the rights of property owners. But we must also note that a similar transformation in the "rights" of labor was taking place. The bargaining power of labor "in the market" was clearly shown to rest on a framework of politics and resultant law, and the political functions of unions thus became as significant to their survival and success as were those called "economic." This became even more clear during World War I when government interest in transportation became more significant than that of either management or labor.

The Operating Brotherhoods took this occasion to press for an eight-hour day and a number of other gains. Acting under the Newlands Act, no agreement was reached. Faced with the threat of a strike, President Wilson succeeded in getting Congress to pass the Adamson Act which forced management to accede to some of labor's demands. The final adjudication of the dispute which led to the threatened strike was made by a special Commission enforcing a decision of the Council of National Defense. The supremacy of the law over the "rights" of management was fully demonstrated. Subsequently, because management was unable or unwilling to maintain a level of operations acceptable to the government, the railroads were taken over in December of 1917 and management was placed in the hands of a Federal Director.

During the period of government operation many of the current institutional arrangements came into existence. Government recognized some of the craft unions developed in other industries, such as the Machinists, unions composed of railroad workers exclusively, such as the Carmen, and of course the Operating unions and the Telegraphers that had already demonstrated on many railroads sufficient power to force their recognition. Since the unions were now to deal with government agencies they took on forms appropriate to these functions. Union structure thus became something quite different than it had been or might have become had the unions developed in response to the acts of private management only.

The Growth of Administrative Tribunals and Law

Of outstanding importance in this respect was the rapid growth of administrative law and its attendant tribunals. It became necessary to spell out in detail what management and labor could respectively do and what they were barred from doing. The interpretation of these rules became more and more crucial to the workers. Administrative decision now determined the lines between crafts, and so helped define the role and status they could occupy if they entered the occupation. Similarly, administrative law set boundaries to the kinds of disciplinary action that management could take. Seniority became the basis for establishing "property rights" in a job. A proposed modification of rules thus came to threaten the career pattern that induced men to enter and remain in railroad service. Because they were so vital, final decisions were no longer to be management's unilateral prerogative. Instead, Boards of Adjustment were created, which were composed of both management and union representatives, and sat to hear appeals from decisions made by management and contested by unions.

The resultant body of precedent, like the common law, was ill defined, at times illogical, and sometimes "irrational." In arriving at decisions the Adjustment Board was enforcing contracts that had been arrived at by bargaining. A particular rule might have been acquiesced to by management in return for labor's acceptance of another condition which they might otherwise have successfully opposed. When this rule, taken out of the context in which it originated, was interpreted by anyone not privy to knowledge about its origin it might appear to be totally unreasonable. With no independent judiciary to reform this kind of contract law derived from collective bargaining, and restate it in terms of general principles, there was a great accumulation of *ad hoc* decisions that defied generalization by the uninitiated. Nevertheless, like the common law itself, it proved to be more effective in keeping the system running than did the more "rational" proposals of those who would substitute statute and code for it. This was borne out by what happened after the emergency of the war was over and the country went back to "normalcy." Government control was ended and management regained some of its power and prerogatives. On the other hand, the power of the unions was greatly lessened. The decisions of government tribunals, once favorable, began to reveal that the new agents followed the election returns. Faced with their adverse judgments, the unions tried to exercise their economic power. In the defeat of the "outlaw" Switchmen's strike and the Shopmen's strike of 1922 their weakness was revealed. The operating unions refused to respect the picket lines and the strikes were lost.

In 1926 the United States Railroad Labor Board, no longer supported by either labor or the carriers, was scuttled. Efforts to re-establish the national Adjustment Boards were defeated. In their place were only some regional or system boards that more nearly reflected the desires of management than labor. With the advent of the New Deal the political climate shifted again. Labor had greater influence with the new administration than with that of President Hoover. Union leaders hoped to re-establish the power they had previously gained and add to it. As in 1920, there was a flurry over the proposal to nationalize the railroads, but no real continuing effort by any group with much political power. The pattern of control that developed for the railroads was somewhat similar to that which characterized the National Recovery Administration, but it took on characteristics that reflected the differences between railroading and other industry.

A federal coordinator was named to reorganize the railroads and bring them into some effective relationship with newly emerging forms of transportation technology. Obviously there were many kinds of interests involved here. To have responded effectively to their varied claims would have been a miraculous achievement. But most of the groups involved either had other interests to pursue or other ways to gain their objectives than to meet the well-entrenched power of the carriers and the unions before the coordinator. They pursued their objectives elsewhere or they were subordinated to the strategy of the two major contenders. The carriers organized and, in part, functioned through the new American Association of Railways and the unions built up the Railway Labor Executives Association, though this did not encompass all of them. The confrontation of these two sets of contenders resulted in a system reflecting the balance of power existing between them. The Railway Labor Act of 1926 was amended to provide the political and legal framework within which railroad labor relations have evolved since that time.

Management escaped many of the controls which once had seemed likely to be imposed on it "in the public interest." Labor got at least a modicum of protection against the arbitrary power of management. In the Washington Job Agreement labor gained compensation for part of the costs workers would bear if proposed mergers went through. They protected themselves to some degree from what appeared to them then as the more dangerous of the results of impending technological change. Once these objectives were obtained they dropped their support of the Office of Coordinator and of general legislation designed to rationalize the whole transportation system of the country. Insofar as I am aware, the Act of 1934 is the only law enacted in the United States that sets up an administrative tribunal which makes decisions in controversies arising out of labor contracts and which are enforceable by the courts. In this respect railroad labor is treated differently from any other kind.

The Railway Adjustment Board set up by the Act is composed of 36 members. Half of these represent the unions and half the railroads. It is divided into four "Divisions," each dealing with the problems of administering contracts with a particular set of unions. The First Division, for example, deals with the five operating unions, and it is by far the busiest. Cases come to these Boards as a result of failure to settle differences between management and labor on the systems they represent. The panel dealing with a particular dispute attempts to interpret an agreement or contract arrived at as a result of previous bargaining. When the panel deadlocks, as is frequent, it is usually along strictly partisan lines, all the members of management being pitted against all those on labor's side. Then a referee must be appointed to join in the making of a decision. Since he is usually unable to convince the proponents of either view that they are wrong he in effect becomes the sole judge of the dispute. It is possible for the members of the Division to agree on a referee, but they seldom do and the National Mediation Board has to name him. Thus the referees are selected by an agency which is part of a particular government in power and reflects its overall philosophy. The decision of these referees is binding on the parties, though they may and occasionally have appealed the decisions to the courts. The influence of politics thus extends down to the interpretation of the rules governing even the most minute detail of the railroader's work and life. The only way the interpretation can be altered is either through the appointment of a different set of referees or through bargaining for a rules change. The former is a long, slow, and uncertain process. The Adjustment Board is not a judicial body which respects *stare decisis.* It is an arena in which nothing is conceded to an opponent until the referee has ruled on each contest. Delay may be as effective as any other tactic and at the moment other than disciplinary cases in the First Division must wait seven years to be heard. In the meantime workers can wonder what will be their fate, and managers guess how much it will cost the company if the rulings they have made are not upheld by the referee.

To resort to the other alternative, bargaining, is not much more productive. Many of the rules now in operation were made when technology was different than it is now and when the power of the unions vis-à-vis the management of a particular road was not the same either. The rules which the unions are defending are often more generous than any that could probably now be negotiated, and they are not about to give them up. The lines between crafts and between seniority districts must be preserved to protect the rights of the members of the various unions. To alter them merely because technology would, if one were starting with a clean slate, justify a different structure than now prevails may seem rational to an outsider. But to the men whose only hold on the economic system is secured through these rules, to consent to such change would appear to be irrational in the extreme. So the unions have few or no concessions to make in return for new gains, and management is unwilling to concede anything without at least a *quid pro quo.*

The upshot is that after bargaining has failed management announces its intent to alter the rules unilaterally and mediation is attempted. That failing, arbitration is proposed, and when that medium is exhausted, the unions give notice that if the rules are changed they will strike. Then an Emergency Board is named by the President of the United States. While its findings are not legally binding, they indicate pretty clearly where the current administration stands. To secure compliance with the findings of the Emergency Board, presidents can use all kinds of pressure, even including threat of seizure of the roads and the drafting of workers into the armed services. So the fiction that collective bargaining (in the usual sense of the meaning given those words) is basic to the agreements administered by the Adjustment Boards finally breaks down and the stark reality of coercion is revealed.

The solution of grievances "on the property" becomes less frequent. Each of the proponents has some idea of where he would stand before an Emergency Board appointed by the president currently in office. Whichever thinks he will get more by using this expedient simply stands fast and waits till the machinery has ground out the specified procedures. In the meantime, day-to-day decisions must be based on rules that are increasingly vital, particularly to the men but in many cases to management also; but at the same time these rules become less and less defensible in terms of changing technology or to the outsider.

Slowly the differential between what management might hope to gain through rules change and what they must currently pay widened to the point that the American Association of Railroads decided to make a frontal attack on them. A prolonged attack on "featherbedding" was made in the press and from the rostrum. This was followed by taking an adamant position in bargaining sessions. The unions fought back with every kind of tactic available to them and the number of disputed rules grew enormously. Finally one of the Emergency Boards recommended the formation of a Presidential Commission to study the rules and make recommendations relative to them. The unions were reluctant to accede, but finally joined in asking for a Commission, which was appointed by President Eisenhower. It was composed of a man from each of the five Operating Brotherhoods, five representing the railroads, and five representing the public. To some who could see the situation of the industry the Commission offered hope for a constructive solution which would revitalize it. There might have been a thoroughgoing analysis of the industry. This would have related labor's position to that of management and the other interested groups. It would have analyzed management's practices to see which were in the public interest. It would have related the technological position of the industry to that of other industries. In short, it would have taken into account all the factors that impinge upon labor-management relations.

Instead, the Commission was limited to an adversary proceeding dealing with only such facts as seemed to have direct bearing on the narrow issues recognized as being in dispute, though there were staff studies showing how some other industries had attempted to deal with somewhat similar situations. In the end the Commission accepted the idea that it should be management's prerogative to initiate most of the changes it wished to make. Labor was in some ways protected. The individual worker was given compensation to offset some of the costs of change. But for the most part the effect of the proposed rule changes would have been to wreck the structure of the unions and set them one against the other.

The outcome was, of course, that the unions refused to accept the decision of the President's Commission as a basis for bargaining. When management in its turn attempted to proceed unilaterally to follow out the Commission's findings the unions turned to the courts and secured injunctions imposing delay. It had in the meantime become apparent that a really effective strike would not be tolerated. Sooner or later there would be a solution dictated by government.

What we will see next is, of course, not entirely clear; the social changes that will emerge as a result of the changing character of technology will be mixed with those arising from many other kinds of change which have also been taking place.

What is clear is that, in spite of broad general ideological support for "private ownership" and "freedom for the worker," technology has continuously altered alternatives in such a way that the social structure emerging resembles only slightly that which would allow the kinds of choices required to serve the "economic man" posited in classical economics. Instead we have a system that reflects the strength of many groups, operating in different arenas, utilizing different kinds of influence to achieve their goals. The outcome is not what would result from choices made in a "free" market, nor the pure influence of ecological variables, nor what would in technological terms be "most efficient." Neither is it a clear reflection of some value orientation. Methods of analysis that will reveal the extent and kinds of influence exerted by each class of variables in this particular situation will help if in turn they can be used to analyze what has taken place elsewhere. We may then learn whether or not certain quite specific sets of social relationships must accompany the use of specific technologies. The task of discovering how a particular system must be altered to achieve that relationship can then be approached with less probable error, and prediction as to whether this kind of change is likely to take place can more accurately be made.

Chapter III

Caliente

In 1949 I made a study of a little desert town which I called Caliente at which was located a railroad division point with roundhouse and repair shops. I set out to see what happened there when the diesel locomotive replaced the steam engine. I need not repeat here the entirety of what I wrote in 1949, but the reader who does not wish to go back to that article[1] may be helped by having at least its outline to understand what I am saying here.

Caliente was created to serve the transcontinental railroad that was then being built between Salt Lake City and Los Angeles. Its location was set by the technological requirements of the steam locomotive, which were such that it had to be serviced at some point in that vicinity—and by local geographic features such as a limited but adequate supply of water, in a canyon which widens at this point as it cuts through the barrier between the Great Basin and the watershed of the Colorado River. From a junction at Caliente a branch line was extended to nearby mining areas that once produced a great deal of wealth but which were, over time,

---

[1]Fred Cottrell, "Death by Dieselization: a Case Study in the Reaction to Technological Change," *American Sociological Review,* 16 (June 1951).

to decline in output. The railroad prolonged the life of these mines by making it cheaper to get the ore to smelters, refineries and markets but it is questionable whether the costs of extending, maintaining and servicing the railroad to Caliente would ever have been met from the income generated by these primary producers. So in a sense Caliente was always dependent for its existence on decisions of men in places remote from it. The decision to build the railroad was made by investors and speculators who were hardly aware of and cared little for the community that was being created as an adjunct to their enterprise. Caliente had value for them only so long as it was necessary to service locomotives, repair cars, and maintain track and signalling systems there. Then, when technology changed so that the demands for such services were reduced or disappeared, Caliente's claims on the outside world declined accordingly. Thus, when the diesel locomotive was put into operation it altered the significance of the geographic environment. It did not need to be serviced at such short intervals as the steam engine, so trains could be run further before they required attention. The railroad company reduced *its* costs by abandoning Caliente as a division point. This wiped out the value of most of the physical structures it had built there. Jobs were eliminated and those who had held them were forced to discover new ones, off the railroad, in or near Caliente, or move to other places where their seniority would guarantee them a railroad job. One of the primary social consequences of this decision was the separation of Caliente residents into sets of people with different life chances. Those with seniority, like the men who operate the trains, used what power they had to take a job at other points. Those who had no property were free to move without suffering the losses incurred by those who owned property. Most of those who had served the railroad employees—the business and professional men—found themselves with heavy investments which they could liquidate only with catastrophic losses. When the immediate results of the closure were concluded only those with a stake in the community, or who saw few life chances elsewhere, were left. So those who were—by middle class norms—most moral were most heavily penalized. Those who had refused, or had been unable, to attain such status, paid a lower price. I predicted a continuing loss of population for Caliente, and assumed that there would be concomitant social disorganization.

In this study I will have to modify those conclusions somewhat. The changed outlook derives in part from the fact that I did not fully understand what was going on at that time. In part it results from changes that have since taken place that I could not then have foreseen.

There does remain a sub-stratum of geographical, ecological, and technical factors which continue to exert influence on the present and future of this little town. We need here to sort the ones whose influence is direct, and unavoidable, from those whose consequences can be and have been modified by variables intervening between them and the people who live there.

This is barren country. The county in which Caliente is located has an area of 10,649 square miles and there are barely more than 2,500 people living in it. It is wild and beautiful but it is extremely demanding of those who stay there. In all that area there are only a few places where man could permanently live without support from outside, and even there with no great certainty. The streams of water necessary for life are few and short and run only intermittently. Most of the sparse rain that falls and the snow that melts evaporates or sinks almost immediately into the soil. Some of this water flows into subterranean pools from which it can be pumped to supply man's needs. But to lift it out of the ground requires a great deal of energy, energy which only recently has become so cheap that crops irrigated with its help can be sold for enough to support the farmer and pay for the pumping. Apart from a few spots then, man lives here only as he can maintain a connection with people outside the desert which induces them to send him the goods and deliver to him the services he requires. In exchange he must either obtain primary products from the earth, or deliver services that are valuable to the people who furnish him with the means necessary to his survival. In many cases the value they place on his product will have as much as or more to do with what happens to him than do his own efforts. It is these anonymous decision-makers who set a value both on what he has to offer and on what he seeks in return for it. Thus it is they who will in large measure determine how many can stay in the desert and how they will live.

In the first study I concentrated primarily on the city of Caliente itself. This put into bold relief what is, in more complex situations, hidden. In many densely populated areas there are a number of different sources of wealth. It is produced in a system that includes numerous institutions that are necessary to production and to the survival of the society in which it takes place. As a consequence it is difficult to determine how much any person or other factor in production, has contributed to their joint product. But both individuals and institutions exert power to secure for themselves a share which they then claim represents their contribution to production, and in turn they consume or distribute this share as their own value priorities indicate they should.

The only contribution of value which Caliente had to offer when it was built was the delivery of the services necessary to keep the railroad running through Caliente. This was paid for in the form of wages and taxes. The amount paid was the resultant of a complex which was determined by, on the one hand, how much the railroad system was able to charge for its services, and on the other, the outcome of conflict between various sets of people, each trying to maximize what they could get as compensation for cooperating in making the railroad run. Obviously we cannot investigate fully all the elements of the complexes that determined either what the railrood took in or those involved in distributing those earnings. But since Caliente itself was so heavily dependent upon this single source of income it is easier to follow the consequences of its reduction than would be true in an environment less demanding than that of the desert.

What Did Happen After Closure?

In this study I have examined not only the city of Caliente but also other communities immediately associated with it in meeting the immediate ecological and economic demands which must be met by Caliente residents. As we have already noted, there were mines and some stock raising as well as subsistence agriculture in this area before the railroad came. The income from these products was exchanged for some of the goods and services produced by Caliente residents. Early in its history Caliente became a shopping center for miners and stockmen, and its hospital took care of many of the sick and injured. But services to Caliente residents were also delivered under the auspices of the county government. Some of these were performed in neighboring communities. In turn these places depended in part on taxes paid by the people of Caliente. To understand what went on in Caliente city we need to enlarge our focus to a county-wide basis. But this picture is still much easier to deal with than would be that of an area with more diverse resources.

The impact of this geographical and biological environment remains very strong. What has altered since I first looked at Caliente is primarily the economic environment. In place of the very few strands that once connected Caliente with the outside world there is a growing network of relationships that influence both what the Caliente resident is able to do and what he tries to do. It also affects what will be the results of his endeavor.

There is little leeway within which the local community can control how its financial resources will be used. Few individuals living there get large incomes which could be seized for redistribution through taxation or claimed by the family, the church, or other means. There *is* general agreement that some people have a right to consume without working. The younger child, the very old, and the greatly handicapped deserve to be supported by diverting income from the channels set for it by the market. But the list is short. All other persons must fend for themselves. Even the families expect that the grown child, if he does not have a job or otherwise contribute financially, will leave the desert and find a way elsewhere to support himself. The effects of these values and attitudes showed up very quickly when the railroad abolished most of the jobs previously located there.

There were no new ones created by the release of job holders into the local economy. It was thus forced to contract. The railroad destroyed or otherwise disposed of most of the physical structures it had built. The residences it had owned were sold off at very low prices to people who expected to remain in the town. This depressed whatever market there might have been for the sale of houses owned by others. Thus much of their value was destroyed even though they remained physically intact. Some of the men who took jobs elsewhere kept their families in Caliente where they could use up the value of their property. They could also keep their children in the schools their taxes were in part paying for. Those who had invested their whole life's earnings in a business which could *not* be moved had to consider how they might survive in Caliente though at a level much lower than that to which they had become accustomed. Those who were unlikely to get a job even if they moved had to estimate what their life chances were to be in a community which was unlikely to justify any great expenditure of local funds in their behalf.

Some of the first effects of the closure were a kind of shocked disbelief. Some people could not really accept the fact that the railroad shops were permanently gone, the jobs eliminated. It was assumed that when management discovered the error of its ways, it would reverse itself. Many railroad families kept their houses with the idea of renting them. This practice was not irrational. The ordinary life of the train and engine men had previously called for them, during the time they were coming up through the ranks, to move away from the place where they owned property. Fluctuations in the demand for railroad service had always resulted in layoffs and furloughs. Shifts in the amount of money spent by the railroad on maintenance of ways and equipment which resulted from managerial decisions based on data totally remote from the ken of most workers, had often occurred, and reorganizations which resulted in shifting the location of various kinds of work were familiar. The worker simply waited for fate to deal a new hand, for the man with more seniority to retire or die, or for new business, generated by war or migration, or shifts in the market, to create the demand for more workers, or for employment resulting from the installation of new equipment or structures to create an opportunity to get back on the railroad which paid most of them far more than they could make in any other occupation. So the psychological results were not immediately what they might have been in a community with a different base. Postponement of purchases of all kinds of durable goods, exhaustion of credit, doubling up of families for the duration, and efforts to create some kind of public works to carry them over a period of unemployment were characteristic ploys. But finally it had to be admitted that Caliente was no longer to be a railroad town. Those left behind as others moved on realized more completely how the survival of each depended on the acts of the others, and values that contributed to the survival of the community began to move up relative to those emphasizing the primacy of private profit.

Even as the economic base of the community shrunk there was active effort to find a new one. There was divided council as to the way resources might be used. Some took a traditional turn. The desert dweller had long depended on local wood for fuel. So a group of townsmen located an area where in the past a forest fire had left the dead trunks of standing pine trees which made excellent firewood. After it was felled and cut they had to bring it down a precipitous canyon, or take a long roundabout, tire and gasoline consuming trip. It was not long before it became clear that, in the absence of the teams of horses that once could be used "free" while they were not needed to till and harvest or to round up cattle, the cost of getting wood for fuel was prohibitive, particularly since few houses were equipped to burn it and the local demand was small. The distance to any market not itself equally able to get firewood nearby made shipping prohibitive. So this form of "regression" to an earlier source proved unsuccessful.

A great many juniper trees grow in the area around Caliente. Many westerners who have camped out are more familiar with the odor and the taste juniper imparts to food cooked over it than they are with the taste and odor of burning hardwood charcoal. Some in Caliente thought that they could develop a market for charcoal chips made from juniper to be used in outdoor cooking. Locally the product sold well. But there was not a large enough base to place and sell this exotic product in the larger cities. Even if demand did increase the crude equipment surviving from early smelting (that first was used to make the charcoal) could not have been cheaply reproduced and those who labored hard and long found that the returns for their work were very low compared with what they had gotten from the railroad, the mines, the highways, and other urban-connected employment. So this venture too failed to create a new base.

But then fate took another turn. The federal government, looking for empty space in which to test and develop atomic devices, found this largely publicly owned and sparsely settled area well suited to their needs. There was no strong resistance to prevent this use of the land. The empty area where the testing was to go on was to be subjected to all kinds of unknown and unknowable risks. Men, women and children in the area wore devices to record how much radioactivity they had been exposed to. And even the jackrabbits were shot and their carcasses analyzed to discover what nuclear pollution was doing to the country. The most important thing to the desert dweller was that this dangerous experiment provided a lot of high-paying jobs. The sons of miners who had been exposed to silicosis and lead poisoning now offered to dig the shafts and tunnels in which atomic devices could be exploded without harm to others. Some of the younger men from Caliente and nearby mining settlements commuted the 320 mile round trip to get the work while living at home. Some of them just pitched a tent and slept on a bedroll for four nights, then drove home for the long week-end. Some "outsiders" who were attracted to the testing area found cheap houses in Caliente and located their families there while they too worked at the test sites. So a few jobs were created, a few children were kept in school, a few customers were served by the stores and service stations, the hospital and the doctor.

Fate further delayed contraction during the Korean War when the primary source of tungsten used in the United States was cut off. The federal government offered very lucrative contracts to those who would find and deliver it. One of the few places where it could be mined profitably is near Caliente. The government paid for a mill and for some housing near the mine, but many of the families remained to use the conveniences found in Caliente itself. This provided a respite. But over time the tungsten mine was closed, and most of the workers at the nuclear test site moved closer to their work as facilities there were improved. Some quit and tried to find jobs nearer to Caliente.

Two other kinds of adaptation to their environment continue to be tried by the people of Caliente. One emphasizes agriculture, the other recreation.

Agriculture as a New Base

There are a few areas in the vicinity where cattle can be grown profitably. This is familiar to the Western pioneers. Whenever there is sufficient ground water to grow grass, hay can be put up to provide winter feed for cattle. Then the beasts can be turned loose to graze on the sparse feed provided on the government-owned open range in the spring and early summer. Cattle are moved up to the highlands as the snow recedes and plants spring up. They are moved back down as the fall comes on, and wintered in the lowlands, feeding on hay. Thus, a limited amount of hay land can support a herd much larger than would be possible if the cattle were pastured or fed from it all year, as is necessary in lands more heavily populated. Caliente itself was founded on one such ranch, but the water and land once used for growing hay became much more valuable as a supply for the railroad engines, shops, and for the needs of people. In areas not more than fifty miles or so from Caliente some new ranches are today being developed as a result of the extension of cheap Hoover Dam power which can pump water to create new crop lands. Big owners with capital, able to take losses for tax purposes, are buying out most of what is privately owned or can be secured from the government. There does not seem to be future employment for more than a handful of people from this source. Raising riding horses for sale or show stock provides a variant of this kind of land use with, again, only a very limited employment opportunity. Most of the land that could be farmed or ranched in the old way is already occupied by the descendants of the pioneers who used every resource they knew about to make a future for their children, most of whom have, even so, had to leave the home place as they matured.

Tourism

The other alternative was to use the country for recreation. Caliente, in common with most western towns, has looked upon tourism as a sort of second class industry. The kind of personality most often developed in a little desert community must be twisted and reshaped before it can fit the servile occupations that deal successfully with the demands of tourists. But there is money in it, and sometimes "beggars can't be choosers." Tourists are attracted actively in a number of ways. Caliente is surrounded by mountains that grow enough browse to feed a fairly large herd of deer. The "crop" can be harvested regularly without damage if hunting is not overdone, for deer are very prolific, but the plants they feed on would quickly be destroyed if the deer were not killed off either by carnivores or guns. Deer season finds Caliente and all the country around it filled with hunters. Many are former residents who come home during the deer season not only to hunt but to renew friendships and family ties. Others are city people who find it invigorating to get into the mountains in the fall, eat and sleep outdoors and throw off some of the effluvia of the city. The deer crop is large enough to keep an increasing number of hunters coming back, but this has necessitated shortening the season, which in turn reduces the income which the businessmen of Caliente can expect to secure from them. So this source too is about as great now as it can be expected to be in the future.

Fishing is another kind of recreation that will attract tourists. In the spring, summer, and early fall, there is fishing in the canyons and the government has increased this by planting fish and putting dams where water can be stored. It is hard to keep these dams from filling with silt, gravel, and stones. Their watersheds support few plants to hold back the water that falls, and erosion is constant. So unless there is recurrent effort to contain the runoff by building upstream dams and dikes to slow it down before it reaches the fishing area, there must be equally constant dredging. Otherwise the dams will fill and their function be destroyed. The cost of this effort is so great that it is highly unlikely that it could be paid for from taxes on the profit the tourists leave with the local residents. If this source is to be maintained, neighboring counties, the state and/or the national government must subsidize these projects for the benefit of fishermen who live elsewhere.

Where the gasoline tax, paid by the tourist, and the state and national government together, pay the cost of building highways, they become net assets to the towns through which they pass. Local businessmen have made considerable efforts to encourage travel on a highway which joins two large national parks and passes through Caliente. One year they sent brochures by first class mail (to prevent their being thrown away unopened!) to a sample of people selected at random from the telephone directory of a large metropolitan area. The results did not encourage further efforts of this kind. A local businessman made motion pictures of the country around Caliente which he showed to service clubs in the cities nearest to it. But the effects did not encourage others to continue in this effort. The highway through Caliente is a main road from parts of Western Canada and the wintering places in Arizona. Caliente has fought to keep this road because some travelers on it stop there and use its motels, hotel, restaurant and filling stations. The old road was put where it would serve early mining and railroad towns. Today road planners are looking for the shortest distance between major cities. The road going through Caliente is not on this shortest route. Towns like Caliente had in the past sufficient representation in the state legislature to exert considerable influence on the road builders. After the United States Supreme Court issued its "one man–one vote" ukase the "cow counties" lost a lot of influence. A new road is to be built which will not go through Caliente.

The federal government is interested in creating opportunities for recreation for the people of the cities and has spent some money to that effect in the Caliente area. It bought a valley long used for pasture and hay production and converted it into a wild life refuge. This however will be on the new road, and its effects will perhaps be to lessen rather than increase economic opportunity in Caliente.

So, in spite of continuous effort to provide a new economic base, the economic picture is bleak. The people of Caliente have continued to believe in the ideas of self-help and enterprise even though this has often meant that they had to turn to the government in their efforts to sustain private business. If we look only at the failure of private enterprises one would expect that contraction would quickly reach the point at which Caliente would no longer be able to keep the minimum number of people to hold a doctor, a hospital, a police department, or perhaps even a school. The decline in population until recently certainly bore out this kind of prediction. Caliente City was incorporated in 1944 with a population just over a thousand. By 1960 it had fallen to 792. But now there are 916 people living there. The population of the 10,649 square mile county has had a parallel history. In 1940, 4,130 people lived there, in 1950, 3,837, in 1960, 2,431, but in 1970 the census showed 2,557.

What About Social Disorganization?

With this kind of economic and demographic contraction one would also expect to find evidence of many kinds of social disorganization, such as divorce, desertion, juvenile delinquency and crime. The facts are, however, that such evidence is hard to discover. Instead we find what is in standard terms a sound community, plagued with fewer problems than most areas where "progress" and affluence are evident. This is so different from the expected that we need perhaps to cite the evidence that it is true before we try to explain why things are the way they are.

There are no empty houses in Caliente, so the familiar use of such places for "immoral" purposes does not exist. There are a half dozen empty stores on the street that fronts the railway rather than the highway. None of their windows are broken, none are boarded up though the shelving and counters and other fixtures are clearly visible within.

The street in front of them is as clean and orderly as in front of the stores still in use. There are a half dozen "saloons" in Caliente. In only one or two are there often more than the bartender and a few customers. They are kept open by owners who have nothing else to do and nowhere to go. They manage to pay their taxes and seldom require police attention.

Caliente has little crime, and since the railroad closed up shop the rate has, if anything, fallen. Most of those who commit crimes are outsiders who are clearly visible and likely to be caught and often part of the sentence they get, if convicted, is the provision that they leave town and stay out. Delinquency is no more prevalent now than in more prosperous times. It fluctuates a little but there is no regular trend. Only one boy has been sent to the state training school since the shops closed. He was so frightened by the hard-core youngsters with whom he was forced to live at the school that he spent most of his time with staff members.

There are a few people receiving local relief. This may result from the fact that the payments made by the town are extremely small and it would be impossible for anybody to survive on them alone. As a result, many of those who might be eligible for welfare payments have had to go elsewhere. The state- and federal-supported Aid to Dependent Children program administered from the county seat has in Caliente about the same proportion of clients as is Caliente's share of the county population. The illegitimacy rate among them is about the same as elsewhere in the state.

So in terms of such indicators as dependency, delinquency or criminality Caliente presents a record far more favorable than do many prosperous and expanding towns.

What About Education?

The schools contribute to the paradox. In 1956 Caliente had 174 children in elementary school (the county high school is not in Caliente). In 1960 there were 159, in 1965 enrollment dropped to 148 but by 1969 it was up to 188. Since the rest of the schools in the county have not shrunk much it is evident that Caliente has not been absorbing many students from other closed schools. The explanation for this situation cannot come from traditional ideas about the way economic contraction affects a community.

The traditional position, that economic contraction will be accompanied by increased disorganization, simply does not fit the picture in Caliente. The economic base *has* shrunk but basic institutions still persist and impart traditional values to the children. In part this has come about through upgrading the values that support the family and the school, at the cost of downgrading other values that in more affluent days seemed important. An example is found in the response to the decline in railroad taxes. The railroad tore down, gave away or abandoned much of its fixed structure. In turn it demanded and got a reappraisal that reduced Caliente's tax revenue.

We can get an insight into the values and attitudes of the people by looking at the way they responded to the loss of railroad taxes. They could simply accept the decline in revenue and cut public expenditures accordingly. Or they could maintain tax revenue by changing the tax rate or by altering the tax appraisals. The first of these alternatives was denied them because city governments in this state may not levy more than one and a half dollars per thousand dollars appraised value on real estate and this was already being done before closure. So Caliente was forced to accept a decline in tax revenue or to increase appraised values. It was evident that in market terms Caliente real estate *has* declined in value. There are a good many evidences of this decline. One was the price put on the "company houses" which the railroad had owned. After giving to the town their water supply and fire fighting equipment, and destroying the things like the roundhouse for which they had no further use, the railroad sold the residences that it had previously rented only to its employees to any who would buy. These brought a very low price. The sale not only evidenced the declining value of all residential property, it also further depressed that price. Houses can be bought today at very low cost. Under "normal" circumstances in most places the reduction in price would sooner or later be accompanied by a reduction in the appraised value for tax purposes. This doesn't take place. Although apart from those not subject to local taxation there have been only two buildings erected since the shop closure, Caliente city has continued to receive almost as much money from local real estate as before. Valuation for tax purposes is shown in table 1.

Table 1. Valuation of Taxable Real Estate

1940	$794,704
1945	$789,299
1950	$765,057
1955	$829,025
1960	$799,809
1965	$786,818
1968	$786,091

Two things should be noted. The first is that even with the decline in their economic position the people elected to retain the appraisals previously put on their property for tax purposes. They were thus *increasing* their taxes relative to the value of their property. The second factor is the fact of inflation, which has greatly *reduced* tax income relative to the *cost* of government. Tax income from real estate would buy in 1969 only forty percent of what it bought in 1940. So even though local people overtaxed themselves they could not continue the services that they previously enjoyed. If these services were to be maintained the income would have to come from elsewhere.

As we pointed out in the first Caliente study the decline in economic opportunity forced a recognition of the significance of values which emphasizes the priority of values contributing to the survival of the community over strictly hedonistic indulgence. This has become more apparent over time. The people in Caliente were intent on supporting their community in spite of market imperatives. Collective effort to secure the ends shared by most community members became more significant as the loss of jobs continued, and the sale of primary products declined. And so in spite of their deep-seated belief in rugged individualism, the people turned to government for help.

New Bases of Community Support

At the frontier, particularly in the desert, the location of governmental services has always been looked upon as a very significant factor affecting local prosperity. The county seat was located in a nearby mining town for some time before Caliente was built. So it could not directly benefit from the jobs county government provided. After the New Deal, federal and state money began to flow in increased amounts through the county, so it was to the state that local legislators turned to provide Caliente with new jobs. With the then current overrepresentation of the "cow counties" in the state legislature Caliente was in a good bargaining position. Its representatives succeeded in getting a girls' training center located there. This resource has had a very stabilizing effect on Caliente's economy, as indicated in table 2.

The school pays a fee of 40,000 dollars to the county for the education of the girls who go to the public high school. It also pays 6,000 dollars to a dentist and 4,800 dollars to the local doctor. It pays a flat fee of 2,400 dollars to the hospital for nonsurgical cases. (There being no surgeon in Caliente such cases are taken to hospitals in other cities.)

Table 2.

Fiscal Period	Average Girls in Residence	Yearly No. Staff	Budget	Payroll
62-63	42	28	379,861	169,292
63-64	62	38	491,919	275,741
64-65	59	38	538,854	327,780
65-66	57	38	611,188	315,292
66-67	53	40	614,151	325,541
67-68*	49	41	588,103	314,782
68-69*	56	43	617,293	333,747

*Note: During the fiscal periods of 1967–68 and 1968–69, the girls' training center entered into a contract with the county school district for educational services. This provided an additional academic staffing consisting of a principal, secretary, orientation director and six teachers.

The budget for 1969–70 was 744,000 dollars of which about sixty percent will be paid out in salaries and wages. Annually about 60,000 dollars worth of the supplies used by the school are purchased locally. The local merchants involved make a profit from the sale of these supplies.

The greatest impact of the center came from the new jobs it created. As of 1969 it employed forty-three people. These include professionals who were educated elsewhere, and a larger number of para-professionals—local people who were trained for the jobs they now hold in special classes conducted by the state government and by the center itself—plus skilled and unskilled labor. It is obvious that this center provides a very important part of a new economic base for Caliente even though it produces nothing for sale in the market and the value of the service it performs cannot be measured there.

Pensions and Social Security

There is another growing source of legitimate income that is increasingly available to people who produce nothing whose worth is measured in the market. These include Social Security payments, veterans' allowances and pensions, railroad retirement annuities, and pensions received by those living in Caliente who receive income from the state retirement system.

For a time it looked as if Caliente was to become a retirement center. Some of the houses which the railroad sold when it closed the shops were bought by people intending to retire there. At first the community welcomed them but enthusiasm waned somewhat as the limited income they brought proved to be low. They were in some cases not able to attain even the spartan level of living the local population had ordained for those not working. Concern over this has diminished since Medicare now prevents the aging from becoming an overwhelming burden on local medical welfare funds. Older people would now be more welcome since Medicare payments help support the doctor and the hospital. But now there are no unoccupied private residences and if vacancies occur they will probably immediately be filled by employees of the center or by people living in nearby ranches who want to move into the town.

In some cases houses are being reclaimed for their own use by railroad men who previously had to leave Caliente in order to get work but now wish to retire in the houses they still own. Their income will in most cases be confined to railroad retirement.

Other Sources of Income

There are a number of federal agencies represented in Caliente. In addition to the post office there are the Bureau of Land Management and the Soil Conservation Service. Around ninety percent of the land in this state is still owned by the federal government. The government, as owner, is represented by the Bureau of Land Management which has had a great deal to say about how public land will be used. It has encouraged the development of grazing and agriculture, and in lesser degree, recreation. It has built a new building in Caliente and maintains a few employees. The Soil Conservation Service, as its name implies, is concerned with practices that reduce destructive erosion through such things as over-grazing, and tries to prevent the introduction of noxious plants. It also maintains an office and a few employees in Caliente.

Other sources of income include Old Age Survivors and Dependents Insurance, Medicare, and Old Age Assistance, which is administered by the state government. The latter has declined in recent years as more of those reaching retirement age become eligible for support from other programs. A good deal of the income deriving from these sources is spent within Caliente.

Another source of income is derived from the educational system. The public schools employ seventy-five people. They constitute the largest employer in the county. Caliente children go to another community when they reach high school. Caliente families thus receive imputed income from the services their children get from this source. On the other hand part of the taxes that support the high school comes *from* Caliente. But most of the income received by the schools comes from either state or federal funds. In 1969 programs sponsored by the latter brought $101,953 into the county. The state government spent $521,136 on the county schools, while a small part of the taxes from which these funds were derived were originally collected in Caliente and the county where it is located. Most of the money came from other sources outside that county. As we indicated above, a very large part of the money spent by the girls' training center came from state or federal government, as did incomes to those who worked on the highways, for the Bureau of Land Management, Soil Erosion Control, the state parks, and Wild Life Refuge.

To summarize, a very small part of Caliente's income is now derived from performing services for those extracting raw material from the earth. The largest part comes through providing services to people in Caliente who do not "earn" their incomes. The remainder comes from delivering services to those who cater to the "dependent" clientele. In Caliente fifty years ago these sets of services would have been important in the reverse order of that which they now occupy. It is in this inversion that we find the reason which permits Caliente to survive.

There are many ghost towns in this region. When they stopped delivering ore to be exported they lost any claim on imports, and they died. Ecology and economy were enough to account for what happened. Apparently there are now means to alter the impact of these factors on communities, else Caliente would have met the decline suffered by many of its erstwhile neighbors. It was expected that the decline would come and would be accompanied by *social* disorganization. We need then to see just why it was thought that social disorganization would accompany economic contraction, if we are to understand why that hasn't happened to Caliente.

Most economists and some sociologists adhere to an exchange theory as the basis for human cooperation and organization. It is assumed that the individual learns by being rewarded when what he does brings satisfaction to others and deprived when he fails to do so. Thus he becomes acculturated or socialized. Those from whom he gets rewards give them only for acts approved in that culture (or in a subculture if he is raised among "deviants"). Somehow a comparison of these inputs and outputs is made by the individual and if this is favorable he continues to act in socially rewarding ways. Some economists use this model particularly in relation to "foreign trade" where boundaries are set so that one can distinguish between those inputs and outputs taking place within the system and those between systems. These economists are often not so much concerned about persistent imbalances within a system as they are about those existing between systems. They account for the value of goods and services exchanged in terms of price, and values that cannot thus be compared are left out of the input-output account. So the market, and price-measured values, are regarded as *the* significant variables to be observed. This model developed for the analysis of foreign trade has frequently come to be used domestically also.

Sociologists, anthropologists, and political scientists have to deal with the whole array of values that humans share if they are to use them in prediction. But many of these factors that affect choice cannot be converted into price-measured quantities. There is endemic conflict over the priority to be given among things that *are not* in the market as well as with those that *are* distributed through price. Nevertheless a good deal of theory rests on the assumption that it is the failure or success of a complex system to deliver *price-measured values* that determines its fate. This is so because values created by older, simpler institutions like the family are not adequate to maintain the elaborate set of relationships which it is necessary to use if one is to make use of complex technology in order to escape the limitations of local ecology. So other values must be subordinated to those that support complex organizations.

In the United States the market and the corporation are presumably thought to be legitimate because they promote and enhance values necessary to technological progress. Many people see these institutions as being less of a threat than government. So in the past if the market ordained that a community be destroyed, that was sufficient reason to wipe it out. On the other hand, the use of the power of the state to preserve a community because it performed other necessary social functions, some of them of a character that the market itself destroys, was thought to be immoral and unnatural.

Under this theory, when Caliente ceased to perform needed functions for the railroad and could not find other means to command goods and services through the market, it should and would die. The reason it did not is of course apparent now. We need to recall that the first response to closure of the shops was selective emigration. We must remember that those left behind as the system contracted were just those most devoted to the preservation of Caliente. They were business and professional men whose life chances could, in their judgment, be better preserved by keeping the town alive than by letting it die. Their way of life depended heavily on such virtues as self-control, delayed gratification, and self-help. They have taught these to their children. They have discouraged the entry of new residents with different ideas. They have prevented the creation of a dependent population and limited the survival of those who were already there through extremely stringent rules pertaining to locally supported welfare. They have in large measure prevented criminal deviation from their norms. They have maintained streets, water and sewer systems, police and fire departments, and partially supported their school system. They do not regard these cooperative efforts as being different from an individual's direct pursuit of his own idiosyncratic ends. These collective efforts are not apart *from* but a part *of* their idea of legitimate activity. A given flow of imports into the community will be and has been used to sustain their whole way of life.

Caliente people assume that, because they are being paid for what they do, it must have economic worth. Not a lot of attention is given to the shift in the structure under which they earn, and within which the value of the kinds of service they perform, is determined. They got money from the railroads for performing acts whose worth they could not judge. It is sufficient for them now that they are paid for performing other services which presumably give satisfaction to others living at a distance from them. They are perhaps more realistic in their assessment of the situation than are those who find it necessary to explain what goes on under a theory that emphasizes reciprocity in exchange.

It is difficult to justify by exchange theory the kind of relationship that keeps Caliente alive today. As we have seen, it still does provide some market-sanctioned services. It houses the traveler, provides a few miners and railroad employees with goods and services and does a few other things that are paid for only because they give satisfaction to those who consume them. But the income of most Caliente people does not come from this kind of activity. Instead it comes from outside sources who pay Caliente people for doing things mostly for each other. They care for the sick. They educate their own children. They work to rehabilitate juvenile girls. They house and provide amenities for older people. They provide parks and other free services for the traveler. They maintain law and order. They keep dependent people from falling below a standard which is set, not in Caliente, but in the state capital or in Washington. So it is only as people elsewhere maintain their own values and the social structure that this requires that Caliente is provided with an income. But with that income it can continue to teach and maintain a set of values that are in many respects contradictory to the values that make the community viable.

We cannot here even suggest how the larger set of priorities is created and maintained. We want only to point out that the system in operation cannot be an "exchange" system unless one arbitrarily decides that nobody contributes anything unless he gets from the source to which he has contributed a *quid pro quo*. Most of those who keep Caliente alive don't even know that it exists, or through what channels their taxes have reached their ultimate destination and been converted into consumption goods and services. They can only receive psychic satisfaction from supporting a total system which they believe will in the long run be, in their judgment, better than one that denies consumption by some people who are not, in the judgment of the market place, entitled to consume. Those who vote for taxes to take care of the old, the young, the sick, the crippled, the disabled, children without adequate parental care, veterans, delinquents, and even criminals certainly consume nothing to which these categories of people have necessarily contributed. It is easier to account for their voting behavior by ascribing it to "ideals" and "ideas" as a source of value judgment than to trace it back to some "consumatory" response: such as is basic to some kinds of psychological theory. To account for their behavior we look at the results of institutional controls that indoctrinate (or "socialize") people so that they feel better when they contribute to the support of the unfortunate among them. Thus the means that Caliente uses to adapt to its ecology run from immediate efforts like digging wells or building dams to the maintenance in other places of religious and moral systems.

Analysis demonstrates that the economic system of the United States is increasingly producing and distributing services (as distinguished from things) as the source of consumer satisfaction. Caliente demonstrates to a perhaps startling degree what can in lesser degree be seen in almost any area of the country. Some of the services the total system provides are distributed by reliance upon values measured in the market place. Increasingly the by-products of the market, such as exhaustion of natural resources, the waste of human beings, pollution, and crime, give rise to group action restraining the unbridled pursuit of profit. People are increasingly becoming aware of the effects that disregard of ecological imperatives can have on the maintenance of a viable society. Those disadvantaged in the market turn to other means to establish new priorities. Caliente reflects these changes. Once it was dependent on decisions made in distant places by railroad stockholders, directors, and managers. Now its fate depends more on the decisions made by voters, many of whom live far away, concerning such things as schools, old age pensions, medical and rehabilitative services. The desert gives it little room in which to maneuver. As a consequence men living there must respond more directly and immediately to the voices of power and influence than do those in most communities. For all the rugged individualism of these men on the frontier, they must listen ever more closely to the beat of a distant drummer to whose cadence they must march.

Tomorrow may provide a different tune. The physical structures that

already exist, the schools, hospitals, stores, streets, and residences provide physical assets which can at the moment be combined with the
income now being sent into the community through government programs to provide services adequate in the eyes of those who control
those programs. As the buildings wear out, as they inevitably must do,
a decision will have to be made as to whether there is any reason why
these or similar structures should be rebuilt. This will force a kind of
quantum jump which may, in the light of then existing social and
governmental structure and the values of the people, be too great to be
justified. Until then it looks as if Caliente will continue to survive
though it produces nothing for export that can be sold in the market
place. But when and if that moment does come, the desert will reclaim
its own.

Part Two

Technology and Aging

Chapter IV

The Technological and Societal Basis of Aging

There is probably nothing on which social scientists agree more completely than upon the thesis that, to a very great extent, social change is tied up with technological change. There is continuous reference to the way modern technology has modified the position of the aged, so in a sense this chapter will refer to the product of interaction between technology and the condition of older people. It will therefore be devoted largely to general theory, while the empirical evidence in its support or contradiction will be found in many of the other chapters.

Technologic Change and Society— Various Typologies

The relationship between technology and other aspects of society has been the subject of a great deal of speculation and has produced a number of theories, at least one of which, Marxism, has led to revolutionary consequences. There has also been accumulated a great deal of empirical data about

the way specific technological changes have resulted in specific alterations in other elements of specific societies. Most of the theories explain
to some degree the observed events for which they are supposed to
account. But none is so well established that postulates derived from the
theory can be shown to be very useful in predicting what *will* occur in
any society into which a given technology is introduced.

This does not mean that these theories have not been useful or effective. In fact, it is often because they anticipate that certain social changes
will result from technological changes that innovators have introduced
new techniques. Decisions to industrialize a society flow from the belief
that this will contribute to the achievement of the goals held by the
decision-makers. Both capitalist and communistic economists agree that
increased productivity, whether measured in terms of price or some
physical unit, is to a great degree the result of using new technology.
Their disagreement lies in a different estimate of the social conditions
under which that technology is likely most rapidly and at least cost to
be introduced. Nor is there agreement as to what other changes are
likely to result when technology *is* changed. It therefore remains an
unfortunate fact that changes in technology still have a great many
consequences which are unforeseen and perhaps unforeseeable by those
who make the decisions which result in technological change.

The best that can be said in behalf of existing theory is that it has
permitted potential innovators to examine some putative consequences
of their acts which pure empiricism would necessarily have neglected.
The potential value of well-developed theory which *would* permit accurate assessment and prediction of the major consequences of technological change is enormous. As a single example, much of the future clearly
depends upon the way we use and control the technology implementing
atomic energy. For this reason every effort at developing theory which
is at all promising should be vigorously tested until its potentiality is
demonstrated. In the meantime, we must make do with such theory as
offers us here the best explanation we know about.

It is obvious that technology has had much to do with the present
condition of older people. Medical technology is responsible for much
of the increased longevity which accounts, in part, for the presence of
so many older people among us. Similarly, developments in nutrition
have led to increased longevity. The substitution of engine power for
muscle power has altered both directly and indirectly the fate of the
aged. Our concern is with the massive sociological changes which have
accompanied technological changes and have also given rise to new
social positions for those in the later years of life. To deal effectively
with these, we will have to specialize in the general and build categories

into which a tremendous number of separate events can be meaningfully classified. . . . Since here we have the task of showing how technology affects all kinds of societies, we shall have to show clearly where the line separating various types is to be drawn, so that, when we get hold of the evidence about a particular people, practicing given techniques, we will know whether to classify them in this category or that one. We could do this, as many have done, by assigning to England the origin of the "industrial revolution" and classifying all countries on the basis of whether they have reached the stage of development which was supposed to have been reached in England at the time of this event. Obviously, this assumes that every society which industrializes will go through the stages England went through, which may be true; but, if we are to characterize these stages, we will have to know just *what* in England's history was essentially due to the technology used and what related to other aspects of English history.

More recently other typologies have come into use. The one which is probably most widespread among American sociologists is that developed by Ogburn in his work *Social Change.* He distinguished between material and non-material culture and held that it is the geometric accumulation of the former which is the cause of technological change. The prestige which this theory holds evidences its usefulness. But there has not been a great deal of empirical research stemming from it, nor has there developed a body of critical literature demonstrating the limitations of the theory which become apparent under careful scrutiny. This lack of interest in a well-substantiated theory concerning the interrelationships among technological and other aspects of societies has deep roots in American culture and is a subject which itself could well occupy all the space at our disposal here, so we will have reluctantly to abandon its further pursuit.

The anthropologists who have concentrated on earlier cultures have made more progress in developing and utilizing a typology. Whether or not their models will serve well to illuminate the modern scene is at least doubtful. As Simmons shows, it is useful to discriminate between the social practices and organizations of herders, food-gatherers, and cultivators. He says, for example, that "self-employment or ancillary services in agrarian systems probably have provided the most secure and continuous occupational status that society at large has yet afforded for the majority of its aged." For comparison, nomadic herders and food-gatherers sometimes have had to go so far as to abandon their aged when survival of younger members demanded it. The results of using this kind of classification are thus shown by Simmons to be of great value in showing some of the social likenesses which are to be found where particular types of technology are used. It also demonstrates that the value systems and the institutions of a society are not completely independent of technological, geographic, and biological facts.

In a recent work, Steward has modified to some degree the classifications of those anthropologists upon whose work Simmons relied most heavily. Instead of classifying cultures in terms of the way sustenance is secured (e.g., by food-gathering, herding, cultivation, etc.), he emphasizes the amount and kinds of energy which can be secured by the techniques which are employed and the way this fact modifies what can be done and what is likely to occur. Steward also revives the use of the evolutionary model but emphasizes its multi-linear character. He considers energy flow to be one of the elements which influences whether or not events will be repeated. Though his theory differs somewhat from that of Simmons and the students upon whom Simmons relies, it seems, on inspection, to fit equally well or better than empirical evidence which is presented in support of the other theories. It has for us the additional value that it can be quantified fairly accurately. Thus, if a new classification of evidence is made concerning the relationships of technology to society, certain figures as to the energy used to do certain things can be worked out for each of the societies to be categorized. If once this has been done, it should become clear that there *is* a direct variation of some social phenomena when the flow of energy is altered; we will have quantified our data and made them susceptible to the use of mathematical and other forms of logic which we otherwise could not use.

It was as a consequence of getting data into this form not only for pre-industrial societies but for all societies that the theory expounded by Cottrell was worked out. The model used, stripped to its skeleton is, like that of Steward, a multi-linear version of evolution. It attempts to show that the survival of patterns of relationships depends in part upon their ability to secure and direct the flow of energy. Chances for survival of any pattern are enhanced if the pattern results in increased flow, and they decline if it impeded that flow. The energy flowing through a particular system can serve through feedback to increase the capacity of that system so that it will secure and channel more energy. So, to use a simple example, a plant, using the energy stored in root or seed, puts out a leaf, which through photosynthesis converts the energy of the sun and the inorganic products of the soil flowing through the leaf into materials which are then built into new roots, stems, and more leaves, which in turn are able to convert more solar energy, and so on, until the limits of the system of which the leaf is a part are reached. Or, to cite another example, food brought overseas by wind-powered ships from a place where there is no shipbuilding timber can be eaten by men in places where there is timber but no food, to build more ships which can deliver more food. The social structure required to permit this kind of relationship among men is just as necessary as are the physical techniques or the materials used. The kind of social structure which permits a particular kind of energy flow to take place is strengthened by feedback from that flow. On the other hand, competing social structures which prevent such a flow are at a disadvantage—a disadvantage which may be very great or quite insignificant but is nonetheless real and worthy of investigation.

The illustrations we have just given are of course not significant as evidence. They are cited only to permit the reader to get a glimpse of the way this basic conception of the way energy exerts an influence upon society is here used. In elaborating the scheme, and providing limited evidence of its usefulness, each of the major sources of energy used by man, such as food and feed, water, wind, coal, petroleum, gas, and nuclear fuel, is studied. This shows how the scientific and technological facts which determine *how* energy from a source used by man sets limits upon and encourages or discourages his use of it for various purposes. Investigated also are means for discovering how much of the energy produced from a given source must be fed back into the system to keep it going and to enlarge it. Thus the amount of "surplus energy" (energy brought under man's control in excess of that previously under his control which was expended to secure it) can be calculated. One of the fundamental theses developed is that it is around the disposal of newly available surplus energy, and around decisions as to whether or not to do what is required to increase the amount of surplus energy available, that much of the conflict in present-day society occurs. As relates to this chapter this thesis would hold that changes in the amount and kinds of energy used in our society constitute one of the most significant factors which have altered the status and role of middle-aged and older people.

Using this model, it is possible to classify societies in terms of the number of units of energy available daily or annually for each member of the population. An arbitrary figure may be set, marking off classes of societies in such terms, and societies distributed into those categories. Continued study of the results so secured may give evidence that societies might most usefully be divided into two or three or five or ten or more categories. But at the moment, and for our purposes here, we have simply divided them into two categories, "low-energy societies" (those in which energy is secured almost entirely from plants and animals, including of course man himself) and "high-energy societies" (those which make use of other sources). As a means of roughly dividing them, a point is set at about 5 horsepower-hours per day per person. Those below that figure are "low-energy" and those above are "high-energy" societies. The reason for making the choice of this figure is that in the limited research done so far we have not found a society utilizing only muscle power which reached anywhere near this figure, while societies using extensive trade through wind-driven ships, such as Rome and Greece and England (before about 1750), apparently did reach somewhere around this figure. Since we regard societies like these latter three as the prototypes of "high-energy" society, we set the lowest figure at a point which would include them.

The Energy Base of Low-Energy Society

The recent development of energy from nuclear fusion and fission has led to a rash of speculation about a brave new world in which most work will be done without human effort. But the energy of the sun, which has always been used by men, is enormous even when compared with that which man can conceivably control through harnessing the atom. It is *not* the absence of a *source* of abundant energy in nature which limits man's use of it. This limit is rather to be found in the process by which that energy can be converted for the purposes to which man seeks to put it.

In low-energy societies this process starts with photosynthesis by plants and ends when man eats those plants to produce mechanical or other forms of energy through his own body or utilizes energy made available to him through the consumption of plants by other animals. Most men in the history of the world, and perhaps even today, live in societies which fall into this category, although almost all men now have access to at least some small use of energy from other sources. Thus all societies, since they all grew up under these conditions, will show some common characteristics which derive from the limits so imposed on them. They may also exhibit other characteristics in which they do not resemble each other. For some purposes the differences between societies sharing the same kind of energy base are far more significant than their likenesses. For other purposes it is the likenesses which matter. If our categorization is useful, it is to be found in dealing with and accounting for the likenesses.

Among the patterns which all low-energy societies will share are those which result from the characteristics of plant life. A given plant can synthesize only a given percentage of the sunlight which falls upon it—no matter how fertile the soil, how abundant the water, how carefully protected from weeds and plant pests it may be. Therefore the people who use it can get only a maximum of so much from that plant. If they are short of land, afflicted by pests, have limited water, and so on, they may not be able to cultivate as many plants as they otherwise could; therefore, they have even less energy available. But, even with abundant land, and without the other handicaps, they, using the energy of their bodies, can cultivate only so much, and the plants will only produce so much. Then, if men want to use the energy of the plant in the form of mechanical energy, they must either eat it or feed it to an animal. The average man can convert only a small amount of food energy into muscle energy per day. The total is only about what an average horse can do in 40 minutes, less than enough each day to keep a 100-watt light bulb burning for 8 hours. So whatever he gets done in work, travel, or play must not in total require more energy than that.

These facts have many social consequences. One of the most significant is that low-energy societies carry on most of their activities in relatively small groups occupying relatively small areas. Until recently, many of their people lived and died without ever leaving the village in which they were born or ever hearing a strange voice or an idea introduced by a stranger. The institutions which such people maintain are those that will serve *their* needs, and these relate intimately to their own survival and that of their immediate relatives and neighbors. They also must assure the continuity of the culture through which the society will be reproduced.

One means of assuring this arises from the fact that all local institutions interact in the same personalities. What is good as defined at work is also good at home, at play, and in worship. The separation of values into such categories as political, economic, religious, and familial is rare. A man's religion supports the family, and the family his religion. There is religious meaning to economic activity, and a portion of what he produces is certain to be devoted to the maintenance of his religion. Most of what is outside the village community and the kinship group is equally foreign and frequently taboo. There is too little energy for many in the group to waste it on extended travel or transport to and from distant places. Such contacts with different people as may take place are likely to be rigorously formalized so as to insulate against change, else they become the source of conflict. Within such limited systems, very extended division of labor, either territorial or functional, is impossible. Most people have to labor at producing food and fiber. Too much labor is required in the cultivation or gathering of food to permit men to spend the time or to reward them for learning difficult and expensive specialties which can be used seldom and only by a few. This reinforces the effects of special limitation for members of a society who not only live close to each other but do pretty much the same things; and, therefore, the consequences of their acts are likely to be pretty much the same as those encountered by their neighbors.

There are few cities, for there is little surplus energy which can be removed from the control of those who produce it without so impoverishing them as to reduce their ability to continue to produce. Thus the bulk of the population remains rural, and the difference between those who live in the cities and those who live on the land is perpetuated so long as the cities survive. Population tends to multiply in the river valleys where surplus energy from food is greater and cheap water transportation can be added to the energy from food.

All these limits affect the kind of communication and social control it is possible to use. Written documents are difficult to reproduce without some form of energy other than man; so spoken language carries the burden of maintaining day-to-day coordination as well as perpetuating the generation. Much social control depends upon tradition spread through the verbal teaching of the elders. We could continue to elaborate the likenesses of low-energy societies, but it would be redundant for our purposes here. Let us rather look at the consequence of these conditions as they affect the character of the institutions to be found in low-energy society.

Institutions in Low-Energy Society

The Family

Perhaps the foremost fact to be noted is the great significance of the family. In most low-energy societies life is relatively short, so that obligation between parents and offspring and the reverse does not often extend to the members of more than two generations. But, in the few cases where grandparents and great-grandparents do survive, their position in the society is usually well fixed, and there is no doubt as to what their roles are or where they stand status wise.

In most of these societies the extended family is characteristic. That is, they recognize obligations and rights among a much greater proportion of those who are "related by blood," one to another, than do the people of the United States generally. In fact, it is only in urban-industrial societies, using very large quantities of energy per capita, that so much emphasis is placed, as it is in America, upon the relation of wife and husband and their children. It is easy to see that, where resources are scarce and productivity limited, the extended family will add greatly to the chances for survival (at least among food cultivators) as compared with the small conjugal family. The flexibility with which the extended family can meet special demands for peak-load energy, such as is required at planting or harvest time, helps greatly to assure the complete use of the resources available. On the other hand, the small family may be unable to use its land or tools effectively because of temporary labor shortage. Similarly, in meeting the catastrophe which comes when a father or mother is made non-productive by accident, illness, or death, the extended family is a superior form. It also provides for a division of labor between the age and sex categories that is more efficient. So, for example, grandmothers and the children can carry on household tasks while the stronger young women do field work; or grandfathers can teach the young boys while their sons are carrying on the more arduous arts. Among low-energy societies, these advantages do not accrue to the small conjugal family.

Let us look for a moment, then, at some of the functions carried on by the family in low-energy societies, keeping in mind that we will later review them to see how they relate to the place of the aged. In the extended family the decisions concerning the birth of children do not rest alone on the husband and wife. It may well be, in some societies, that no girl will be accepted into her husband's family until she has borne a child or even a male child. Later, unless she produces her share of the offspring, she is not considered to be a good wife and will suffer the consequences. Unless a man has sons, he may be denied a share in the lands held by his family, since his daughters have no such claim and will be forced to marry into another family; or the reverse may be true, and rights to use land or other property may descend only through daughters. On the other hand, where long experience has taught that the land will produce only a given amount of food, excess children may be unwanted. The survival of a new baby may here be assured only by starving others now living. Thus the group may require abortion or infanticide or may adopt taboos in relation to sexual intercourse which effectively control the birth rate. We do not wish to go further into this, generalizing on the kinds of controls and their extent; we want only to point out that in the extended family the control over the birth rate lies at a point different from that in the conjugal small family system. The decision-makers are not the same in the two systems, nor are the results to them of limiting births or permitting unlimited births the same.

In the conjugal family there are, to the parents, many predictable consequents of childlessness. These extend from such facts as the extinction of a property-inheriting direct family line to being deprived of the affections of children and inability to provide for one's own emergencies and catastrophes. On the other hand, in an extended family the inheritance of property by one's own children may not be thought of as being particularly significant, and a man may receive affection and care from nieces and nephews much like that shared in the small family only between parents and children. Thus, to the husband and wife, the consequences of bearing children may be quite different in the two systems.

Romantic marriage, widely prevalent where "individualism" is considered to be the ultimate test of social well-being, encourages a relation in which those who decide whether or not to beget and rear children are expected to make the judgment largely in terms of the effect this decision will have on them personally. But the effects of the birth rate extend to a great many people who are thus put in a position where they are unable to influence the choice being made. The aged are probably the most significantly affected group of all these, and in recent times they have probably become the most numerous.

In low-energy society the family carries on the greater part of the economic activities performed there. Among the cultivators of food the labor force generally consists of the members of the family physically capable of engaging in production, though often there are also work groups made up from the members of different families. Since the division of labor between the sexes is largely based on the fact that a man is physically not capable of doing some of the things necessary in the reproduction and the early feeding of infants, women do what is not incompatible with their role as biological mothers. There is a great variety of roles which are played by men in one society and by women in another, but some things women cannot do if they are to reproduce and suckle the young child. Therefore, food preparation and as much of its production as is compatible with woman's necessary role are likely to be carried on by women. Activities demanding heavier musculature are carried on by men. Here, again, what we are emphasizing is that decisions as to what is to be produced, by whom, where, when, and how, are made within the family and kept at least minimally compatible with the primary function of the family. This fact has important implications for aging and the aged.

Within its capabilities the family also provides the means to assure the care of the sick and injured and to maintain the health of its members. As we have seen, most available economic resources are in the hands of those who have a concern for the well-being of the family. It is usually also true that the obligation to care for family members who are unable to care for themselves is fixed in the cultures, so even if the members of a particular family do not personally care about what happens to one of them, they are constrained to look out for him because of the sanctions which will be imposed upon them by the rest of the community if they do not do so.

The family also serves to pass on to its members much of the culture which they receive. In the process, of course, the values which are necessary to the survival of the society are created. The most significant of these—the sacred or religious elements—usually include values necessary to the preservation of the family itself. Like other elements of the sacred, "Honor thy father and thy mother" is not merely a slogan to be repeated on auspicious occasions but a directive, the violation of which may have dire consequences, including sometimes banishment, torture, or death. Sometimes the wrath of the gods falls on those who fail to respect their elders.

The family in this case, and very frequently where supernatural sanction is not often used to secure compliance with mundane obligations, may exercise physical coercion. The head of the family may be held accountable for the behavior of any of its members, but he is, in turn, legitimately entitled to punish them physically and to deprive them of privileges.

In consequence of all these facts and many which we cannot elaborate here, the family is in a position to dictate many of the roles which an individual may play and the status which is attached to them and hence to him. The conception that he has any being apart from the family may be completely absent; but, even where there is recognition of the person apart from the role, the range of autonomous behavior is greatly circumscribed. From birth to death the individual retains his identity in the family, which is itself immortal. The idea that at some time in his life he would cease to be associated with his children and their children may be as much or more difficult to conceive or to accept as is cannibalism or incest. This again has very significant consequences for the elders, which we will examine as we consider the aged in the total environment of low-energy society.

Religion and the Church

Among people in low-energy societies that which is sacred tends to be served by special persons designated to carry on the functions necessary to preserve and implement it. In some cases the head of a family may himself be so designated, and "ancestor worship," which sanctifies family ways and makes saints out of real or mythical forebears, prevails. In many more, there is a kind of priesthood which functions among the membership of a number of families. It is their task to assure that the sacred elements of the culture are passed on unchanged to the next generation. They must also sanction acts prescribed or proscribed. Much of what is sacred behavior takes place in or through the family, and the structure which maintains the sacred is very likely to be devoted to a great extent to the preservation of family life. Thus "the church," as we in the West call it, supports a morality which glorifies and sanctifies the human family, and the family in turn teaches children, when they are very young, the necessity to respect and honor the church.

In some low-energy societies the church has itself become an economic system which competes with other forms. Very often this competing economic system is devoted to the care of those who are not cared for through failure to function of another economic system such as the family. Care of the unattached sick, the blind, the mentally ill and retarded, the epileptic, crippled, and the aged has frequently been maintained out of surplus energy derived from the land by childless servants of the church. So long as agriculture remained the prime source of energy, and the converters used consisted primarily of human beings, an institution such as this was well fitted to compete with or supplement the family or the manor. Old technology and the social structure capable of serving it were sanctioned through time. Sacred ways and technologically competent ways were not necessarily antithetic.

Government

As contrasted with the very great significance of family and religious values, low-energy society makes much less use of government than does that using more energy. We here designate government as the institution which performs a complex or cluster of functions centered around or dependent upon the legitimate use of physical coercion. Many low-energy societies assign the roles required to carry on such functions to the members of the family and the religious institution, and there is no secular state at all. But many others, particularly those which developed out of irrigated agriculture, developed an institution in which the power to coerce was monopolized. That is to say, they might permit the church or the family to coerce its own membership but denied them the right to coerce others. At the same time the agents of the state became legitimately entitled to coerce within limits both church and family members.

Where the governmental institution emerged, as separated from the family and church, it performed only limited functions. It was not expected that it would operate as a complete social system. Apart from its police function to control the behavior of certain types of deviants, the most common activity was that of making war or providing for the common defense. Whether the police function resulted from the necessity to create a means to control large amounts of energy to be used in conquest or its repulsion, or whether military potential resulted from the aggregation of force necessary for "internal" policing, the fact that such an aggregation *was* created and maintained is crucial for understanding the functions of government. We do not, here, have the time to dwell on the various theories of the way the state originated. We want merely to point out that the power legitimately to exercise force in a particular manner is related to the whole moral system of the society and, hence, to its total social structure.

Political power, to be legitimate, rests fundamentally upon what children learn in the family, the church, the neighborhood, and other "character-forming" experiences. In low-energy society neither the family nor the church was willing or able to abdicate all its functions in favor of an omni-competent state. Rulers were thus confined in the functions they controlled to those which did not have the effect of disrupting or destroying the substructure upon which the society rested. Such of them as disregarded this necessity were able only to destroy the society; they could not build a self-perpetuating total state. No state in low-energy society could seize absolute power, though it might well seize almost all the surplus energy produced in that society—and some of them did and do. Apart from the police and military functions of the state, what, then, are the functions it commonly performs in low-energy society?

A very common one is that of protecting "property." The amount and kinds of things which may be classed as such vary tremendously among societies. As we shall see later, a great deal of what is classed this way today did not exist in low-energy society; thus the state could have no power over it. On the other hand, for example, property in slaves, fairly common in low-energy society, is rare in high-energy society.

In low-energy society there was frequently no idea at all of the personal possession of land as such. A particular territory was recognized as "belonging" to a tribe, clan, or other social entity. A particular area was conceived to be that on which a particular family was expected to make its living or from which it was entitled to take natural products. Even sections of the sea might be claimed in this way. But there was no thought that this right could be alienated or permanently transmitted to another person not sharing the role in which the property was attached as a prerogative. The state protected a common right of a group, a family, a continuously existing entity, not that of an individual. So, to select an example from our own system, a particular President of the United States is not expected to be able to sell the White House or dispose of its furnishings by giving them to his favorites or deeding them to members of his family. Our present concept of property "owned" by a corporation, like Harvard University, corresponds more nearly to the idea of ownership of land in most low-energy societies than does the idea of property which one individual can dispose of as he sees fit.

In low-energy society, where land is basic to the production of energy, the power of the state to transfer title to land is usually limited; and the irresponsible exercise of this power is very likely to rob the state of its legitimacy. The power of the state is thus quite small as compared with that which it might have where other competing energy systems exist to provide a different basis for power than that which lies in the hands of the food cultivators.

Very often in low-energy society there is little resort to the state to resolve personal disputes or to secure redress for injury. Morality may well require the individual to seek to deal with problems of this kind or find their solution through resort to the family head or the priest. The formation and dissolution of marriages may be considered to be matters of no concern to the state, even though they are of great concern to the church; and the sanction of family feuds may be the basis for solving family disputes. Thus many decisions about family obligations lie outside the competence of secular governors.

On the other hand, where sacred elements of the culture call for physical coercion, the state may be required to enforce the sanctity of the marriage or to permit family members to violate its monopoly on the use of coercion by resort to the "unwritten law." In this case the state is more the servant of the family than its master. Similarly, the secular state may be required to protect the prerogatives of the priest, on pain of having its claims to legitimacy reduced or destroyed. The norms are thus set, not through political machinery such as legislature or a bureaucracy, but through social processes in which heads of families and the priesthood play the dominant roles. These are likely to be the elders.

In such low-energy societies as have developed spatially extensive political systems, and in particular in irrigated areas, the state also protects communication. To keep the flow of water at the proper level, in the proper place and time, requires constant accurate communication. Similarly, the mobilization of force for military and police purposes requires that the governor's messages are conveyed accurately and swiftly. A communication network necessary to perform a function, like the control over the water of a river, may be utilized to extend the military power of a centralized government. This, for example, was certainly one element of the rise of such government in China. But here, again, we must recognize that government was controlling only one very limited type of communication among people, most of whose thoughts were transmitted locally and verbally, so that control over communication was distinctly limited.

Similarly, in areas subject to periodic crop failure, the existence of a central agency which has the power to seize surpluses produced in good years and hold them for lean ones may contribute to survival sufficiently to compensate for the cost of maintaining governmental services. But here, again, it must be recalled that these surpluses are created by people in local, family-community-type economic systems who are likely to resist passively, if not actively, unless they accept the legitimacy of the state in this activity. In the few cities possible to low-energy society, the state occasionally has also been expected to carry on services like the provision of sewers, canals, roads, and aqueducts; but these functions pertain to the limited commercial or industrial areas much more than to agricultural places.

Only extremely rarely is "government" expected to provide for the welfare functions so common to it in high-energy society. On occasion a tribal leader might attempt to alleviate temporarily the condition of a village or family hit by catastrophe. Some particular form of dependency may be considered to be related to the sacred in such a way that special care is given to those suffering from it. On the whole, however, responsibility for the care of the ill, the injured, the blind or deformed, the mentally diseased or deficient, and the aged falls upon the family or, in feudal society, upon the manor, with an occasional share devolving upon institutionalized religion. It is from such societies as this that much of our present culture stems; it is not surprising that it has been difficult for people reared in this culture to accept government as legitimately carrying on these functions.

The Market

Of all the institutions which characterize America today, none is more ubiquitous than the market. But, on the other hand, in low-energy society the role of the "free market" is extremely circumscribed. As we have seen, the great bulk of economic activity took place under the control of the family and community, with an assist from organized religion and limited participation by government. The idea that the economic activities of men are separable from their familial, religious,

and moral or political consequences was unthinkable under these circumstances. The disadvantage of trading where there was to be no
regard for the consequences of economic production and consumption
other than those felt by the trader at the moment when he made a trade
was quite apparent. Until recently there was no system of production
which made use of the market, which was also so much more physically
productive than that carried on in traditional ways that it could compensate for these quite apparent liabilities. Thus, such trade as was
permitted at all was usually confined to a very limited area and to goods
not necessary for the preservation of the sacred culture. Recently, some
of the proponents of business civilization have attempted to show that
the "natural" form of society is one in which all man's values are put
on the auction block to be ordered on the basis of his economic desires
as implemented by his wealth. Such a view will not bear more than a
moment's scrutiny in the light of man's history. The widespread importance of the market as an institution and of market mentality is very
recent and relatively rare in the history of the peoples of the world,
though it has always played much larger and more significant functions
in the urban region of low-energy society than in the food-producing
regions outside the city. It was not often that the aged found themselves
entirely dependent on the market for life's necessities.

There are a good many other elements of low-energy society that
affected the role of the aged. It is not our purpose here to show more
than the barest outlines of its institutional character, in the most general
of terms. We do it so that we may see more clearly the position of the
aged in the societies from which the American people came and in such
portions of American society as retained the way of life dependent upon
the exploitation of energy initially converted by plants. So now let us
look at the position of the aged in light of what we have learned.

Position of the Aged in
Low-Energy Society

Control over Family and Population

As we pointed out, it was the extended family which determined the
birth rate. This put the aged in a strategic location to influence it. Every
man who lives long enough will grow old. As he looks at the contemporary condition of the aged, he contemplates his own future. The addition of another child to the family has, to grandfather, much the same
meaning as it has for father, who will be grandfather tomorrow. If a man

begets too many children, there will not be sufficient food to nourish all of them and provide also for him. If too few, there will be no one to care for him in his old age. It does not require excessive rationality to observe this relationship; and, where the particular individual is thoughtless, the culture is likely to carry sanctions against his deviancy from what is considered moral in this respect. Control over birth may be ineffective because the vagaries of life are such as to provide no clear guide as to what the number of children should be; epidemic disease, for example, may occur so irregularly that no system of planning can assure the "proper" number of births to maintain the "proper" numbers. Similarly, the very early death of most people may make it desirable that every possible child be born. But, for the most part, low-energy societies have had societal controls which served to preserve a balance between births and deaths well inside those which would have resulted from complete dependency upon starvation and disease to bring about that balance. Thus, it was usually not necessary purposefully to limit the lives of the aged.

It is not true that people in every one of the cultures which contributed to the American civilization had an equally high regard for the aged or gave them the dominant voice in the determination of the birth rate. But, in most of them, the voice of the elders, expressed in the culture and transmitted by them to the children, was likely to be listened to much more attentively than it is in America today.

Control of Economic and Social Roles

The influence of the aged in this, as well as many other respects, was greatly enhanced by the control they exercised over economic activities. Family life has been so idealized in recent years in the West that its exploitive character is often neglected. If we see it clearly, however, we may realize how fully one age or sex group may exploit another where the family is absolutely necessary to individual well-being and survival. Men do not live long in low-energy society; thus the concept "aged" should be used carefully. But the head of the family was frequently the oldest surviving member of the group. He had much to do with the assignment of tasks and the reward to be secured for performance, although he was, of course, greatly circumscribed in his control by tradition and custom. But, as anyone who has ever worked knows, there are a myriad of ways by which a work supervisor, regardless of what rules he is supposed to work under, can favor one worker over another. The wishes of the old man were often influential when he exercised economic control, and his favor was particularly sought after where the culture permitted him to determine who would inherit goods, position, or power. Similarly, the old woman could determine a good many of the consequences which would flow from the acts of the younger ones; and, where the sexual division of labor limited women to very few roles, almost all of them centered on child-bearing and care, the rule of the old woman might be even more absolute than that of the old man. Care of the aged was among the first of the obligations laid on family members. The strong were required to care for the weak, the ill, and the unfavored; and, since the aged often were in a position to decide whether they or some other dependent should be favored, they occupied here, too, a strategic position.

His frequent right, as family head, to ordain physical punishment to members of the group gave the elder an effective means to secure his own well-being. It did not apparently occur to the young Isaac that he should resist when his aging father, Abraham, decided he should be offered as a sacrifice to Jehovah. The Jews were not atypical in their attitude that the old man's word was law, no matter how harsh his edict might seem to his offspring. Their position could be duplicated in many low-energy societies.

But power did not stop with the exercise of coercion; it could be extended through supernatural sanction too. Violation of taboo, as it related to the position of the aged, might mean eternal damnation as well as mundane punishment. Secure in his ability to wield this weapon, the old man could envisage his physical decline without fear that it meant loss of power.

He then controlled tremendous resources, not the least of which was that of assigning within limits role and status to individual members of the society and thus forming a personality and a social structure which would inevitably assure that he would not become a non-entity by loss of function and control. He could see to it that what the child learned would produce an image of the aged, such that it would require destruction of his whole personality to extirpate it or even change it much. The variations in societal behavior which threatened his position in each generation were dropped out when cultural transmission took place, unless there was some compelling reason why succeeding generations should repeat them; but the idealized image of the aged was transmitted over and over again. Where a particular family head was absent, or unable or unwilling to insist upon his prerogatives, both state and church often insisted upon them. Judges, both secular and sacred, were most often themselves elders. A man too weak to demand of his children the honor due him might be despised or even punished. The offspring who failed to give respect due the aged were even more likely to feel the pressure of the organized state or church. In the administration of governmental services, too, it was often the aged who played the dominant roles. The picture of old men sending young ones to their death in war has not faded even in modern times. The tests of battle, however, did put a limit on the capacity of the old to rule in military society or during wars. Most of the fighting has to be done by younger men, and the recurrent rise of the corporal to seize the marshal's baton is good evidence that where low-energy society exploited young men too harshly it might be altered or destroyed by them.

The use of forced levies to build roads, aqueducts, or canals also had the effect of weakening communal, family, and religious controls over the young. Where extensive public works were developed, these forces were likely to constitute a check on family and community dominance. This was particularly true where the right to cultivate the land descended to only one of the sons, the others being left to attach themselves to the economic system in some other role than that of husbandman. Frequently, they joined an army which had to be paid for "protection." No matter what the economic system, it is rare that the young males are starved for long, though they may be otherwise greatly underprivileged.

Influence of Low-Energy System Cultures on the United States

The United States, like all other modern states, was built up by people whose culture was derived from low-energy systems. A great many of its institutions can be understood only by tracing the origin of its present characteristics back to situations much like those of which we have been speaking. The Polish family might differ greatly from that of the English, the German from the Italian, the Japanese from the Chinese; but all shared in greater or less degree the characteristics which were necessary to the culture, social structure, and the values of these people while they depended upon plants for their energy.

As Simmons has indicated, there was no difference among pre-industrial societies which is as great in degree as the differences which separate them from industrial societies. Whether we trace a particular aspect of our cultural heritage back to the Continent or to Pre-Columbian America, we find elements which reflect the nature of low-energy society.

The basic religions of the United States emphasize the primary values developed in the family. They idealize family relationships and regard the Deity as being one whose characteristics symbolize the father figure. All make sacred some forms of family living. All taboo some kinds of relationships which might in fact better serve high-energy technology than does the inherited sacred structure. The conditions of life encountered in pre-industrial societies have had much to do with creating the matrix of institutions, morals, and religion which make up a large part of present-day American culture.

Some Elements of High-Energy Technology

As we have indicated, we are here separating those societies which make use of less than 5 horsepower-hours of energy per person per day from those which use more. Obviously, any technology has many attributes besides its ability to channel energy. A typology based on those elements may in some cases serve better than that which we use here. Students of our economic system have paid considerable attention to the advantages of scale and those of subdivision of labor. Both, of course, contribute to the effectiveness with which energy is used. It is possible to produce a great deal more, using the same amount of energy, if men are highly trained in a specialty and co-ordinated by a specialist in management, than if each man has to do many things at some of which he is likely not to be so competent as in others. It is possible so to specialize, however, only if there is a relatively large demand for identical goods from the same source. Hence large-scale production, standardized consumption, and specialization go hand in hand. These social attributes may characterize some operations even in low-energy society, but the costs of transportation by muscle power prevent widespread trade and thus limit the size of the market. Low-cost transport is thus needed to make gains from specialization and scale possible. Such transport depends upon the use of energy not secured by way of plants as converters. Thus, while neither specialization nor scale owe their gains directly to a new form of energy, they indirectly depend on its use. Of course it is also true that many societies have existed in which there was available far more energy than could be used, given the social organization and other elements of the society operating there. Gains made only as a result of superior technique and organization often result in increased use of energy so that it serves rather as an index of their effectiveness than as a cause in itself. Our emphasis upon energy is, however, directed at another element of the system, that is, the dynamic effect of surplus energy.

The major source of surplus energy used in the world until around 1750 was that secured from plants. Since these plants have fixed limits on the amount of energy they can convert, and since until very recently it was not possible deliberately to alter them much, surplus energy was always small and relatively constant. The use of the steam engine to capture surplus energy stored in fossil fuels made it possible to increase energy from these sources at a rate totally disproportionate to that which man had previously experienced. The addition of the internal-combustion engine and the electrical generator made surplus from coal, gas, petroleum, and, now, the atom available for thousands of purposes. The outcome, of course, is seen in the proliferation of machines to convert energy from these sources, together with gross changes in the necessary social organization, in the growth, size, location of population, and in values.

Table 1.   Growth of Horsepower-Hours of Energy Produced and of Population in the United States, 1850–1950

Year	Horsepower Hours (In Billions)	Population (In Millions)	Horsepower-Hours (Per Person)
1850	10	23.2	440
1900	78	76.0	1,030
1950	675	150.7	4,470

Source: C. Tibbitts, "Aging as a Modern Social Achievement," in C. Tibbitts and Wilma Donahue (eds.), *Aging in Today's Society* (Englewood Cliffs, N.J.: Prentice-Hall, Inc., 1960).

In Table 1, Tibbitts has set forth the changes which have taken place in the amount of energy used in the United States. He has also demonstrated how this increase in energy has been accompanied by a change in the source of energy and in the converter through which it was put to use (Table 2).

More recent figures show that the energy output per person is still climbing though perhaps at a less accelerated rate than previously. Obviously, too, the percentage of energy derived from human beings is beginning to approach so small a figure that even tremendous increases in the energy used will not reduce it much.

Also of outstanding importance is the increased length of our working years. . . . Remaining life-expectancy for males at age 20 has increased from 42.2 in 1900 and a work life of 39.4 in that year to a life-expectancy of 69.5 years in 1955 and a work life of 43.0 years. The productive years available to society from each person born into it are thus on the average greater than was formerly the case. We might also deal at length with the way variance in the demographic profile has accompanied increased use of energy. But it would be extremely difficult if not impossible to indicate the degree to which these changes are dependent upon technological as distinguished from other changes. This is likewise true with regard to such alterations as the amount of urbanization and industrialization taking place, each of which is shown ... to have very significantly altered the meaning of aging in the modern world. Here we can only take up broadly the consequences of change as they are related to the use of new fuels and their converters.

Table 2.   Percentage Distribution of Sources of Energy Used in Producing
Goods and Services, 1850–1950

Year	All Sources	Human	Animal	Inanimate
1850	100	13	52	35
1900	100	5	22	73
1950	100	1	1	98

Source: C. Tibbits, "Aging as a Modern Social Achievement," in C. Tibbitts and Wilma Donahue (eds.), *Aging in Today's Society* (Englewood Cliffs, N.J.: Prentice-Hall, Inc., 1960).

The first social consequence in the use of high-energy technology in point of time and perhaps also in terms of significance was the destruction of the integrity of the village community as the basic unit of cultures based primarily upon agriculture. Outstanding in this process was the divorce of consumption and production patterns from the land on which they take place. It will be recalled that, where cultivated plants furnished the bulk of energy surpluses, there was necessarily some accommodation at the local level between those who consume and those who produce. Wind power operating on the sailing ship permitted goods produced at points quite remote from one another to be exchanged. The social codes under which production took place were therefore not related to those under which consumption of those goods took place in the intimate manner which had characterized production and consumption in low-energy society. This terminated the automatic character of a great deal of social adjustment. The processes of time alone could not certainly be relied upon to produce reciprocal adjustments. Those in control of transportation by ship could redirect the flow of goods as their self-interest dictated. This interest was rarely equated with that of the consumers or the producers, so that, when changed conditions might have righted an imbalance, the trader was able to respond to them in a way quite unlike that which was possible to a producer rooted through soil to the local situation. For example, he could introduce into a market food produced by slave labor and with it support the growth of cities which would have stopped growing had they depended upon local produce. He could continue this process so that the local food-raiser was reduced to penury, but the moral outcry of the farmer went unheard in the market places of the city. City-reared soldiers fed with cheap or free slave-produced food could be depended upon to put down any rebellion among the tillers of the soil. The gradual destruction of the local husbandry, the erosion of land, and the destruction of the old social organization might go unnoted by those fed from ships. While the use of the price system to mediate the relationship between producer and consumer thus stripped both acts of much of the moral significance they once had, it still remained true that the moral consequences of both production and consumption constituted an integral part of the social system of any society. We will later note in greater detail what some of these consequences were. We here want only to emphasize that it was the fact that the new techniques could use great quantities of free surplus energy which led to a great part of the changed character of economic acts. Those who adopted these techniques gained in economic power. They simultaneously gained political and military power. Those who refused to permit the kind of trade which made possible the use of increased surplus thus became more and more vulnerable before those wielding the power of new technology.

As we have indicated, low-energy society rested on a land base composed for the most part of village communities or of small groups moving about in quite narrowly limited geographic areas. High-energy technology undermines this kind of unit.

Low-energy societies which are based on muscles as prime movers cannot escape the limitation imposed on them by the necessity to use land for the production of food and feed. But, when they shift to the use of, say, petroleum products and means to convert them into work, the nature of the limits change, and so does the kind of production which is least costly and most productive. The human is much more often paid in proportion to the time it takes to do a job than in proportion to the energy expended upon it. When human energy was the basic source of energy, however, there was little opportunity to increase the energy used without simultaneously increasing the time taken. This is where the new fuels became so important. Those who produce petroleum products make available for man's use thousands of times the energy which is used to produce them. Thus a farmer can, by increasing the energy he uses, greatly speed up the operations he performs; but, since the surplus energy from petroleum is produced at so little human time cost, the time gained by using it is greatly in excess of that spent in producing the surplus energy. As a matter of fact, in the United States we have reached the point where more calories are expended in the production of some crops than can be secured from those crops. It is apparent that those who control production in agriculture find it increasingly advantageous to use techniques requiring more and more energy from petroleum and other cheap sources, substituting it for time payments to human beings. Since a man driving a tractor can cultivate or harvest a much larger area in a given time than one can who drives a draft animal or uses hand power, many fewer people are needed in the old cultivated areas. They must, if they are to use the new techniques effectively, migrate to places where they too can have access to fuel-produced power. So the village community which is the backbone of low-energy society is threatened both by the decline in its effectiveness to serve agriculture and by the development of the market. In its place there arises a new kind of rural life which is often served almost as poorly by the old institutions as is the urban industrial complex into which so large a part of the population is drawn by the character of the technology it serves.

It would not be possible in this short space to delve into all the relationships between changing technology, changing values, and other ecological factors which give rise to and affect the character of urban life. We here point only to the part that the modern metropolitan community could not exist without the power grid, steam and internal-combustion engines, and the fuels which serve them. And one other thing is certain: there is no high-energy society in which cities are not growing. An ordering of societies in terms of their energy consumption places them in almost the same position as does ordering them in terms of their urbanization. The city and high-energy technology go hand in hand. But city life is not well served by the institutions which developed in low-energy society.

The new technology also required a complete change in the nature of the division of labor. Age and sex grading remain universal forms of social structure which have very important consequences for the aged. But the assignment of roles and statuses in terms of those functions that are required by new techniques are of perhaps equal consequence. The body of science and technology which is in use has grown to such enormous proportions that no man can know more than a tiny fraction of it. Effectively to use it, some new kinds of social organization are required. The knowledge once possessed only by the elders is not enough. In many cases it is, in fact, obsolete, and its possession is a hindrance rather than a help.

To co-ordinate the activities of a great many people, each a specialist in terms of his knowledge and skills, requires more than the family or similar kinship organization can provide. A science and an art of management are needed and new institutions to serve them. The concern of management for efficient use of resources replaces the concern of the traditionalist for the preservation of the old social structure, at whatever cost in terms of efficiency. Members of functional groups, exercising power and influence as such, place demands upon the system in which values associated with the preservation of the family, the community, and the state are sometimes subordinated to those of the union, the corporation, or the profession. Since in many cases these organizations serve as the nexus through which most of the major values of the individual are secured to him, values which enhance the power and primacy of that organization gain for him ascendancy over those necessary for the preservation of older forms of organization. The fractionization of morality is thus in some degree a direct result of technology itself, and in others it derives from the emergence of new forms of social organization which grow up to serve technology.

New means of communication have also appeared. Through the mass media it is possible for outsiders acting at a distance successfully to compete in the promotion of values with those who rely for their success upon face-to-face communication in primary groups. Since it was the monopoly upon child-rearing which gave the family in the local community so strategic a place in low-energy society, the destruction of that monopoly by the establishment of other institutional controls carries with it a threat to all family-and-community-centered value hierarchies. The development of new educational and promotional organizations has been affected to an as yet unknown degree by modern technology, but the arts of mass communication have been much more frequently put at the service of the market, the corporation, and the state than the family or the community. Reliance upon the family to transmit the culture has declined as the character of the culture has made such reliance an impossibility. To an increasing degree education has become a function of government and of specialized research and information agencies. While it remains true that a great reliance is still placed on the family to provide the primary-group values upon which so much of the rest of the social structure depends, the family is not capable of providing the technical information or the specialized morality necessary to the performance of the modern roles of the individual. Technology has thus served at almost every turn to weaken and destroy the patterns familiar to low-energy society.

What must happen, then, as high-energy technology develops is a reconsideration of the values and the institutions of the society into which it is spreading or within which it is evolving. Every value which affected choice in the old system will be reexamined in the light of its costs in the new one. Every institution will be under strain, losing or gaining in its capacity to perform functions for the members of the society in which it is operating. In this welter of change the position of the aged will be altered along with that of every other category of persons.

We may see more clearly just how the aged are affected by noting the specific changes which occur among institutions by seeing what the status of the aged is in these institutions and thus discovering how the status of the aged is altered by these changes. We may also note how the shift from community relations in rural society to those found to exist in urban society also has given rise to alterations in the life-chances of the older person.

Changing Position of Institutions in High-Energy Society

Decline of the Extended Family's Power and Functions

As we have already indicated, much of the old structure depended for its effectiveness upon the fact that men in a local community commonly shared many of the same tasks, faced the same conditions, and fared the same way under given local conditions. But, because the use of high-energy technology calls for extremes of specialization, the extended family is a very poor device to provide effective social structure. The division of labor between the sexes and age groups may, in a given specialty, be necessarily quite different from that which results from the formation and production of family life. The family cannot produce the specialists needed. The skills are transmitted outside the family. Specialized knowledge must be acquired elsewhere. Industrial discipline may require morals quite different from those that could be made compatible with the requirements for family stability. Rarely is the family large enough to consume the total output of a production unit or to use high-energy technology effectively in production. For example, efforts to utilize only the members of a single family to conduct such coordinated activities as are carried on by the New York Central, Standard Oil, or United States Steel must obviously fail. With the failure of the family to serve technology goes loss of its ability to assign role and determine status. With it, also, goes the ability effectively to transmit control over the economic means to living through social inheritance under family control. The old man thus loses a great deal of his power. His prerogatives decline in numbers and significance. He is no longer the teacher; he cannot initiate youth into the mysteries of effective knowledge and skill under conditions which will assure that, at the same time, youth will be required to absorb and effectively adopt codes which protect the weakening aged from the energy of the young. In fact, the very skills and knowledge upon which he depended to assure his control may become obsolescent and thus impose a handicap upon the pious youngster who heeds the old man's advice or respects his position.

In place of the extended family appears the small conjugal system based primarily on the marital relationship. This type of group can much more easily meet the demands for rapid mobility, both spatial and social, which are imposed by changing technology, than can the extended family which served low-energy society so well. The old man no longer selects his daughters-in-law; nor can a mother, through influencing a choice of groom, assure her future position. In fact, it is difficult to convince the young that the family has any legitimate function after the children have selected mates of their own. It is difficult to impose on them a sense of obligation to their elders. Claims established by the state, the corporation, the union, and particularly by their own children take priority. Those priorities are respected by the parents, who can expect to enter no effective protests when, in the interest of "getting ahead," their children enter employment far from their parent's place of residence in a world whose codes are bound to alienate them from their parents.

As we have observed, the exploitation of high-energy technology requires this redistribution of population at an ever increasing rate. The proportion of the populace residing outside the metropolitan areas continuously declines, and the emergence of new population centers follows from the invention of new means of production, new materials, and new weapons, vehicles, machines, and other industrial products. So the assertion of old rights in the name of piety, and the demonstration that effective personal living depends upon family unity, the persistence of family ties, and the maintenance of childhood friendships, is met by the conflicting demands of the state, the corporation, and the market—demands implemented by the flow of energy from sources which the extended family cannot effectively control. With the decline of the extended family goes a loss of control by the aged over the birth rate. The young adult is free to choose how many children he will have, solely on the basis of the values which he holds at the moment. He may be without the experience to understand how an increase in his children will rob him of his ability to aid his parents or even provide for his old age. He may be unmoved by the fact that, in enlarging his family, he renders housing inadequate, capital insufficient, and the number of hospitals, roads, and other collective services incapable of serving his own and future generations. Whatever influences him during those few years that he is begetting children is alone considered to be of significance. It is no longer in the hands of the aged to affect the factor, which probably, more than any single thing, will affect them.

As already indicated, in place of face-to-face communication, there is mass communication or specialized publics, each learning elements of life not shared with others. Neither permits the development through interaction of an effective common code. The persistence of ideal images of the aged, once assured, now becomes extremely doubtful. In the market place of ideologies such symbols as carry the father image of the past compete with those new symbols dreamed up by the businessmen and the proponents of power politics.

Secularized knowledge is required for the effective use of modern technology. It conflicts at a myriad of points with the sacred. To assure technological competence, even the church is required to teach this body of knowledge which undermines its own authority. Continuous "reinterpretation" of the gospel creates schisms and throws doubt on the validity of the whole theology. With the undermining of supernatural authority, the old man is deprived of another means to assure his position. There is no longer a bastion to protect him against the superior energy and knowledge of his young competitor.

The decline in the size of the group which can claim the services of its stronger and wiser members for the protection of the weaker, sick, or otherwise ineffective members means that the family can no longer assure its members against such disasters. It also means that the position of the old man in the family gives him no strategic advantage over others who are unable successfully to compete in the struggle for life. The great emphasis upon youth, which a rapidly changing society like ours is apt to adopt, is accompanied by increasing concern over all the unfortunate young as against the declining aged. Public education of the young is justified by the contribution they may yet make. Worry over the young deviant is multiplied by the years of damage he may yet inflict upon society. The dramatization of the young victim of polio or palsy is much easier than it is presenting an appealing picture of the problems of the aged. Aid to dependent children fits the pattern of responsibility to one's children which is generated in the conjugal family. The reciprocal claims of parents on their children are in this context often considered by parents themselves to border on the immoral. Since the claims of all the dependent are no longer mediated by the elders in a family system, the decision-makers today are less likely to see, understand, and respect the needs of the aged.

The necessity to set up special institutions to teach secular knowledge and technological skills has given the state a tremendous advantage over the family and local community in the transmission of culture. In fact, it has permitted the creation for the first time of a system in which the young may systematically be taught to despise the ways of their forebears. The United States is a nation composed of immigrants and of their offspring. Many of them came bearing cultures which were not only in conflict with each other but almost always poorly suited to provide for living in a country continually advancing to new frontiers, both geographic and technological. To "Americanize" the child almost always meant, and frequently means, that family ways are "old-fashioned" or "old country." In either case they were such that, if the child attempted to persist in them at school, he was likely to be ridiculed by the teacher and most often by his peers as well. Rapid changes in technology and in science have made it dangerous for the youngster to depend upon his parent for help in school. In part, this results from advancing the frontiers of substantive knowledge, and part of it derives from the development of new techniques of teaching which the parents neither understand nor approve. When the new methods are combined with the traditional means known to parents, the result is confusing to the child who wants to know the "right" way to solve his problems.

The Church

The aged can no longer rest secure in the belief that other institutions will reinforce their authority. They can, in fact, be almost certain of at least some conflict with other authorities which may serve to weaken not only their own power but also that of all other agencies so that the child may become alienated from all legitimate control. This great increase in secular knowledge not only reduces the role and authority of the aged in the family but also reduces the power and the authority of the priesthood. Of all the agencies creating moral behavior, the church probably has demonstrated more continuous care for the aged than any other, except the family itself. But its capacity to do so is weakened by the fact that, while it can verbally sanctify the love of the aged and glorify their care, the decisions as to what physical goods shall be used in their behalf no longer can be made primarily among the elders of the church, for it can no longer supply the necessary goods from its own economic resources. Beginning with the revolutions which accompanied the kinds of trading which the widespread use of the sailing ship made possible and profitable, the position of the Western church as an economic system was progressively undermined. Most of its lands were taken from it or rendered relatively less economically productive than they had previously been. So the religious agencies, too, have lost much of their capacity to serve those for whom the family is unable or unwilling to care. Stripped of its wealth-getting activities, the Western church is relatively less well able to provide for the aged pious than in earlier times. Thus neither of the institutions which once provided a fairly powerful position, from which the aged could promote and administer to their own needs, can any longer be depended upon adequately to do so.

Legal Institutions

Nor do the aged fare much better when they rely upon the dictates of ancient law administered by the elders. One of the early means through which adjustment was made to trade and industrial production was the substitution of private contract for public law. Under it "free men," in return for a place in the new economic system, were permitted to barter away most of the protections built into the old by agreeing to forego what the law had previously guaranteed to them. The courts were then required, in most instances, to enforce contractual obligations even though this might have the effect of making many other parts of the old social system inoperable. The employer was absolved from legal responsibilities toward his aging employee by terminating the contract. It is now clear that it is not possible this easily to dissolve the social obligations which a breadwinner or a citizen is expected to meet. Yet the courts recognized no legal obligations establishing the rights of the individual to such economic resources as would permit him to do what morals and law required him to do. Thus the aged were barred from asking, in the name of an old system of law, what changing social institutions denied them.

Moreover, it is not only in terms of this substitution of contract for status that the legal position of the aged has been altered. With the rise of high-energy technology there have emerged new forms of social control almost, if not quite, as effective and pervasive as the common or civil law. Administrative law and industrial jurisprudence grow out of the special requirement of high-energy technology and the bureaucracy which is required to serve it. Professional codes and union rules provide a whole range of effective controls which lie outside the bounds of what is dealt with by legislatures and courts. Occasionally, these codes are modified, or the sphere they are permitted to rule is reduced by statute. Again the courts, operating on the basis that they exceed the limits permitted by the Constitution, reduce the sphere within which these controls operate. For the most part, however, they are allowed to function with only the barest kind of interference because the rules in question have grown up in very special situations. A high degree of technological competence is required to understand and interpret them —a kind of competence seldom shared by the old men who occupy the bench and know only the traditional forms of law. Sometimes the elders, among those who participate actively in the evolution of these specialized codes, are able to use them to reassure a position for themselves. Union seniority provisions represent a case in point. But far more frequently the changing demands of the situation make technological competence an absolute necessity and functional efficiency of greater moment than the preservation of inherited social structure. They thus deny any great influence to those who may be interested in the preservation of social structure and of human values not immediately pertaining to the problems confronting the administrative decision-makers.

Increasing Functions of the Market

As we have already indicated, the use of price-measured value stripped from its connection with other kinds of value is not widespread in the history of man. It was the introduction of new technology which served to upset the way goods and services were exchanged in low-energy societies and permitted them to be offered freely in the market place. We need, perhaps, to look for a moment or two at the reasons why this took place before proceeding to show what has been its significance to institutional development.

We have already pointed out how, in low-energy society, most of man's activity tends to be centered around a very limited geographic site. Here, the costs of goods and services can fairly easily be ascertained. The number of bushels a husbandman produces and the number of man-days which go into their production often remain relatively constant even over generations. Similarly, almost everyone knows the number of sheep required to provide the wool, for, say a coat, the number of days the spinner must spin, the time it takes the weaver to weave it, and the difficulty of learning and performing the necessary skills. Over time, since all goods and services depend upon the same energy sources, and all use man as a common converter, a system of rewards and deprivations gets worked out which is adequate to secure, generation after generation, the behavior necessary to produce and distribute the goods and services which a people utilize. This system may be quite exploitive of one set of people and reward another quite "unfairly," as compared with what some egalitarian or other moral principle would dictate. It may, for example, provide rewards primarily on the basis of assigned status, with little attention being given to "productivity" as measured some other way. The point is not that this stable arrangement conforms to some universal principle of justice or of technical efficiency but that it becomes capable of survival because it is part of what successively induces generations of people to do what they must do to make the system work.

Once any such system gets working, it is difficult to alter it. The culture, the social structure, and the value system serve to assure that each class of persons necessary for production will be taught to carry on its necessary activities. They will also learn that variation from their assigned roles will be considered to be immoral and may result in severe punishment, including in extreme cases even eternal damnation. The "just price" promoted in medieval Europe is only one example of this kind of system. In low-energy society generally, then, goods and services are produced in a social matrix which largely establishes their rate of exchange. Often the economic function is intimately connected with another; thus, for example, families may agree to an interchange of goods at a rate fixed through the intermarriage of their members. This arrangement will, in many cases, continue until that marriage is dissolved. It is easy to see that this kind of relationship is much easier to maintain when the limits under which production takes place are permanent and do not fluctuate widely from time to time. There is, then, little opportunity continuously to alter physical productivity in any particular direction so that, say, the weaver is able to produce more and more cloth in a given time.

The introduction of a new source of energy changes this relationship. As noted earlier with the use of the wind-driven ship, it was possible to bring goods from afar, produced there in some kind of stable system which made their exchange value low, to other points in which it was higher. From there in turn the ship could move goods with low exchange value to other points where they would bring a higher return.

What must be remembered is that the exchange value of goods in low-energy society was based only in part on physical productivity. Thus, a trader might, in exchanging goods between two societies, be taking advantage of the fact that the role and status system of one society differed from another in the rewards it offered—a given productive agent. For example, the weaver in one society, because he was the head of the family, might be rewarded more highly for a day's work than was the spinner, who in that society might be an unmarried sister of the breadwinner or of his wife. She, having fewer social responsibilities and lower status, could expect only less for her day's work. In another society perhaps the situation would be different, and the obligations and the rewards of the spinner would be as great as, or greater than, those of the weaver. An exchange which permitted the trader, and perhaps the consumer, to benefit from lower exchange value of goods produced by one kind of worker in each of these systems would have the effect of upsetting both systems without necessarily producing an increase in their physical productivity. The trader could encourage such an exchange and seek the support of such consumers as were more concerned about the exchange value of cloth than other values which were sacrificed to permit trade. But the original producers who suffered from this kind of exchange would similarly seek support from all those who looked with suspicion or fear upon alteration of the sacred ways or who discovered the ramifications which resulted from, say, depriving the weaver of the means adequately to play his expected role as breadwinner.

The introduction of trade, then, might greatly upset the whole system of statuses and roles upon which the operation of a society rested. Those who wished to preserve their sacred system could be expected to react to prevent trade, and since, in the circumstance cited above, the trading system produced no real increase in the goods and services available, expansion of trade might be limited or even stopped altogether.

On the other hand, the introduction of the ship, permitting trade at a distance, sometimes resulted in real, persistent gains in physical output by some groups in the economy. So, for example, it permitted more wool to be grown in areas where wool, because of climate and soil, could be produced more abundantly than food and exchanged with those living in an area where the reverse was true. Thus both food and wool became more abundant among the trading partners. While this kind of shift in land use and in the roles which producers were required to play might also have disastrous consequences to the previous system, the rewards offered for compliance with the demands of technology were sufficiently high, for example, to turn many British corn fields into sheep runs. This real increase in physical production could be used to encourage or force enlargement of exchange based on price. Thus the introduction of the new energy source, which required in turn the enlargement of the market, also affected the way the family or other institutions operated. If to preserve its old ways one society prevented trade, they also prevented the use of the ship and the surplus energy it could provide them. It was often subsequently overrun by a neighboring one which, complying with the demands of technology, had made it possible to support a bigger population, accumulate more wealth, or otherwise increase its power. The range of the market was then further extended, and gains which arose from specialization of labor and from larger scale production were added to those arising originally from ecological differences.

It was often not possible to find culturally sanctioned exchange relationships extending between people who did not share the same culture. Under these circumstances a system of free markets developed rapidly, money became a common denominator of values, and price served to reflect the alternatives which different societies permitted among their members.

With the use of the price system, goods could be secured on the basis of their desirability as objects without reference to the conditions under which they were produced. Exchange value in the market, measured in price terms, thus replaced the values sanctioned by the community, its culture, and its social structure.

But, in the market, values which could not be dealt with in price terms were frequently made subordinate to price considerations. Of course, where this new ordering of values was sufficiently disturbing, there was reaction which in its turn limited the use of pricing. But in the West there was no turning back. Price-measured values rose higher and higher in the hierarchy of values.

With the price system established to carry on some kinds of exchange, the opportunity to add other energy sources was expanded. To supplement the energy of the weaver and spinner, the British added that of falling water and eventually coal. The skills and the social organization required to supply an increasing demand for cloth were supported by many groups which could benefit by this physically more effective production system. The prestige and power of those associated with trade, shipbuilding, and textiles were raised. It is not necessary here to show how British industry designed to serve trade expanded from this base, so that eventually most of the energy used was coming not from plants, wind, or water power but from coal. Nor can we explore the way other nations, who borrowed British technology and set it down in their own society, modified the British model.

The point of particular interest to us is the theory which developed around British experience, for it is that theory which had so much to do with the way institutions have developed in the United States. As we saw, in British experience it was through the use of the price system that trade was expanded, and it was through trade via the sailing ship that a new source of energy was added to those generally used in low-energy society. Given British conditions, the physical gains in productivity were inseparable from the particular kind of social structure which facilitates trade. Thus the use of the price system and the gains in productivity which resulted from substituting coal for plants as a source of energy were inextricably tied together. British theorists could not see how an industrial system could emerge or be sustained without the free market, by which they had themselves escaped the barriers to economic productivity established by low-energy societies. They thus held that the only way great gains in physical productivity might be made was through permitting the market and the values and social structure which it supports and which support it to take precedence over such other values and structure as would interfere with it. Government, for example, could create nothing, and, if it were to be effective, it could interfere only with a system which would in the long run assign every person to the role he should take and provide him the "proper" reward for his productivity. On the other hand, since "you can't interfere with the law of supply and demand," government itself would eventually be forced to recognize the claims of the market; there was no reason, therefore, for it to intervene at all.

This was their verbal position; but a study of British history will reveal that the trader was himself continually using political and military power to expand the area in which he was to be free to operate. The British Empire was not created merely by allowing British salesmen to peddle the cheaper products of British industry to an eager world. The power of government was constantly being used to prevent those who were opposed to the spread of trade and the disturbances in their value system which it brought with it from interfering. Freedom of the seas meant that British traders could go where they pleased, but, so long as Britannia ruled the waves, there was no such guaranty to other traders. In fact, then, it is apparent that it was, perhaps, as much because of the *political power* of the trader and the industrialist who was allied with him that Britain became the workshop of the world as because of the *price system* which they also used to achieve this position. It will be recalled by those who know British history that "Free Trade," as a doctrine, appeared only after British industry had become so physically productive that it need no longer interpose political and military barriers to protect it against other systems.

Be that as it may, the doctrine that price-measured values are dominant among those shared by men became an article of faith among American businessmen, and, insofar as the territory of the United States itself was concerned, this doctrine was implemented by means which prevented the various states from interfering with interstate commerce. While undermining the social structure of low-energy society, businessmen led the movement which reduced the capacity of the states to

protect their citizens against the ravages of disorganization which accompanied the dominant use of high-energy technology served by the
free market.

Position of the Aged in the Free Market

We cannot explore this thesis further, but it has been developed
sufficiently far that we can now examine the position in which it put
the aged. We have seen that the family and religious institutions were
stripped of many of their most important functions through their inability to direct and control the new forms of technology. How, then,
do the aged fare in the free market?

The first aspect to be noted is that the market tends to penalize those
who permit values other than those measured in the market place to
interfere with its operation. If the purpose of those in the market is to
secure for themselves the greatest price-measured value, they will seldom find that this result is secured by taking first into account the
integrity of the family, the advantage to the state, or gains to the church.
Thus, for example, the banker who is permitted only to measure in
monetary terms the consequences of his handling of the bank's money
may not refuse to collect from the widow or foreclose the mortgage on
the orphan's heritage. To do so would be to violate his obligation to
those who expect that their money will be secure and earn all that it is
possible to collect. This kind of situation is one which the defenders of
a pure price system have some difficulties with. They waver between
attempting to justify the collection of all that the market will bear by
showing that "in the long run" this is certain to produce the best of all
possible economic worlds, and, on the other hand, denying that the
price system in action is in fact so heartless as it is made to appear. It
is clear, however, that if managers permit other considerations than
price to prevail in the market, they violate the propositions upon which
the proponents of the free market base their case.

This case rests on the assumption that no other kind of value judgment is as fair or as economically sound as that secured by freely
offering goods for exchange in the market. So nothing justified the
interposition of any other kind of value judgment.

This point was not made with the idea of invading the field of economic theory. It was designed to show that, by the logic of the market,
there is no place for consideration of many of the values by which the
aged were in the past able to buttress their competitive position against

others. The use of the labor market puts a penalty on aging, as will be shown over and over. It does so because, in market terms, the aged are frequently not worth as much as younger men. Many people suffer in aging a decline in the abilities which the market is willing to use. It is often easier to train the young than to retrain the old. The young have yet more years of top productivity remaining to them. On the average, they have more physical stamina and greater agility and dexterity. They are not burdened with internalized standards of production which have been made obsolete by the use of power-driven machinery. Where these things count, and often they do count in determining price-measured productivity, the market penalizes the employer who hires or continues to employ the aged. The market is also ruthless in selecting for extinction the firms which are not, in terms of the values measured in the market, as efficient as their competitors. Thus the savings of the aging proprietor may go down the drain with the chances for future employment of his older employees. Where a business does not fail, its securities may greatly decline in value, so that aging investors become the victim of the market-measured inefficiency of the firm in which they have invested their savings.

There is no place in the market for the unemployable aged, widowed, disabled, or mentally incompetent. But the market itself is frequently responsible for having destroyed the economic capacity of some other agency, like the family, to care for those whose services have no price-measured value.

In the market the young unencumbered worker and others with few social responsibilities receive a reward based on their market-measured productivity. So, too, do those with many social obligations. In effect, then, the irresponsible are given higher rewards than those who carry on more socially desirable activities. The market treats human labor as if it were like any other commodity to be bought and sold. But the worker who accommodates himself to the system of rewards and punishments which the market establishes frequently must become immoral or dysfunctional in terms of other roles he plays. The investor finds that technological obsolescence threatens the security of his investment as competitors, unburdened by plant and machinery adapted to an earlier method of production, are able to offer goods at a price which he cannot meet and simultaneously pay for outmoded equipment and obsolescent man power.

Thus for all the flexibility which the market permits, it has distinct handicaps as a means to organize economic activity. Each type of person affected by these adverse effects has tried to assure its own position. Investors and managers were the first to be able to do this.

Development of the
Corporation and Its Effect
on the Position of the Aged

Early captains of industry found in the corporation an institution which was able to provide a new framework in which the new technology could be operated successfully but which at the same time extended much greater security to the investor than he could secure in the market. Government was called upon to create this legal person. In addition, it provided patents, trade marks, tariffs, and subsidies which the corporate manager was able to use in such a way that he could escape many of the dangers of free competition in the market place. Thus those of the aged who were protected by this new economic institution found themselves more secure. But those who remained in the free market faced new insecurities. The free market spreads the danger of loss among all those involved in a transaction. But the corporation which protects its investors and managers simultaneously throws an additional burden upon others who are left to keep the necessary balance between supply and demand in the market. Aging farmers, workers, and small proprietors thus gained little security through the rise of the corporation and the administered price system which has developed to serve it.

Until quite recently corporate managers regarded the protection of their stockholders and maximization of gains to stockholders and managers as their sole concern, and most still make all judgments with an eye to the largest profit that can safely be made. They continued to use the market mechanism to determine the value of the services of "outsiders" even though they had to develop a different means to evaluate the services of corporate "insiders." For this reason the corporation was no better a means than the market itself to serve the needs of unorganized and unprotected workers, small businessmen, and "unemployables," among whom, of course, there were numerous older people.

So far, then, the emergence of high-energy technology had been accompanied by weakening of the elements of the social structure upon which the aged had in the past relied to secure their ends. The new emphasis upon the market and other economic institutions, while it made the system more technically competent to provide an increased flow of goods, provided no certain way by which the aged could claim a share of that increase or even to guarantee to them the economic goods and services which they were able to enjoy in low-energy society. As we have indicated, they were not alone in this respect. Unorganized workers of all ages found themselves in the same condition, as did small farmers, the widowed and orphaned, the sick, injured, and otherwise handicapped. Under these circumstances we should expect that they would all simultaneously be seeking means through which their well-being could more certainly be secured.

The history of the United States bears witness to the variety of movements that were spawned in this situation. There is not room here for more than a passing glance at the way the different needs of various sets of people gave rise to specific movements. There have, however, been a host of such experiments. Many, such as the Amish, for example, sought to re-create the closed community based on self-subsistent agriculture. They were willing to forego the gains to be made by adopting technological change in order to preserve traditional values and institutions. Others, accepting the new technology, sought to channel its gains differently from what would have been done by the operations of the free market or those administered-price systems which were favored by managers and investors. The trade unions and the farmers set out to gain for themselves what semimonopolistic trade was delivering to other groups. Very commonly, the farmers made use of their votes in the legislature to direct the state in such a way as to benefit them.

Their action took the form of establishing private title to land at very low cost, providing credit at a rate far below what was ordained by the money market, the creation of scientific and technological research, the results of which were distributed freely to farmers, and, more recently, direct controls designed to give the farmer "parity" with industrial groups. The institutions thus developed did not provide any particular gain to the aged as such and, in fact, in many cases may have served to deprive them of some benefits they might otherwise have reaped.

The same has generally been true of other new organizations such as the trade unions. While seniority provisions have helped to assure positions to older members of some craft unions, it has also happened particularly in the industrial unions that the agreements made have required the early retirement of capable older workers to make way for younger union men.

During the great depression many of these disadvantaged groups were combined in support of the New Deal. It was only then that older people as such got favored treatment. In part this probably stemmed from their growing potential political power. In part it represented the effort of other age groups to take over the roles of their elders without at the same time violating basic moral values. In part it stemmed from the recognition of the justice of the claim put in by those in the later years of their lives. . . .

Current Status of the Aged

There are, however, a few generalizations about the position of the aged in the community which we might beneficially note here. There is a general decline in the power position of the aged. Studies indicate that growing patterns of retirement result in the forced abdication of the holder of even very powerful positions in the hierarchy of management. As Rose indicates, there is a decline in participation of those over 54 years of age in voluntary organizations. A lower percentage of those over 54 vote than in the 10-year cohort just younger than they. There are a few elective offices to which a man once elected is likely to be continually re-elected so long as he chooses to run, and in these older people predominate. However, the functions of the organizations controlled by those in the later years are declining in significance, and the power they wield is correspondingly less.

Activities in which there has been an increased participation by the elders greater than their proportional increase in the population are of less public and more personal concern, like recreation and activities connected with physical health and its maintenance. But there is not much evidence that this increased participation by older people in these activities has been accompanied by growing power over them. There are few communities in which either the health or recreation facilities and activities are designed primarily to serve the needs of older people. What is done is as likely to be done for them as by them.

Research to discover what are the changes in the public image of the aged is not currently conclusive. As Williams shows, the people of the United States do not generally share the prevalent European attitude that old people should be let alone to enjoy life as they see fit. This is not the image the aged have of themselves, nor is it that which others hold. Further research may show how increased leisure time and other fruits of increased productivity are being distributed to the various age groups and the way this is likely to affect the changing image of the elders.

In the meantime conflicting efforts to create an image are being undertaken. On the one hand, there is the effort to portray the latter years of life as the "golden years"; on the other are groups showing the same period of life to be one full of deprivation and unnecessary anxiety. Those who would succeed in gaining public support for subsidies with which to relieve the condition of older people must present a picture showing how dire is the need for these subsidies. Since the effort must be directed at the entire age category, differences in the condition of various sets of aging people are disregarded. This is, of course, most true in the pursuit of social legislation. Where specific groups suffering from specific ailments or handicaps are involved, the differences may be emphasized. However, *were* the private agencies which seek to teach people to anticipate the later years rather than to dread them *to be* successful in portraying the image they pursue, public support for those presumably now living well and happily would undoubtedly fall off. So efforts to create one image reflecting one status are at least in part offset by efforts to demonstrate the existence of another. . . .

The degree to which change is attributable to technology is in some cases quite manifest. In others technology seems to be only remotely related to what is going on. While it would be interesting to tease out the strands of the complexes which lead back to technology, it would in many cases be almost useless to do so because the ability to change other factors more manifestly and immediately connected with what is happening provides more strategic and efficient means to achieve desired goals.

On the other hand, many "reform" movements, particularly those which are in fact reactionary, are doomed from the start by reason of the fact that they do not take into account recalcitrant technological facts. Often they would interfere with or prevent the adoption of technological changes manifestly beneficial to other age groups and often to some sets of older people themselves. Those who would benefit take action designed to permit them to make the desired technological changes, and the conflict which results is far more frequently resolved by making technological changes beneficial to others than to the aged. Thus those who seek to alter the condition of older people need carefully to scrutinize the way the changes they propose will affect technology. By studying intimately the outcome of various proposals for action in some terms not colored by values themselves (such as energy, which was proposed here), it may be possible to avoid commitment to programs proscribed by their technological effects. Often there are a number of paths leading to the same goal. If those ways, precluded by their technological effects, are abandoned in favor of other means which do not bear this handicap, the ends sought will more certainly be achieved and at less cost. For this reason, an examination of the technological origins and consequences of various social movements may well be worth more careful attention than has previously been given them. In any case it is apparent that the discovery of technological components of a social system is of more than historical interest. It is one of the basic foundations of any effective action program.

Part Three

Technology and Peace

Chapter V

Men
Cry
Peace

The very means by which men have intermittently sought peace throughout recorded history have recurrently resulted in war. Such disparity between ends sought and results obtained is certainly undesirable. It is precisely in the reduction of the disparity between ends sought and results obtained that science has been most helpful in the past. This is the domain of science. Just how spectacular the reduction in disparity must be before the method is called *science* is not yet a matter of common agreement, but in any event the scientific method is justified in that it avoids more error than does any other means to knowledge. Whether or not a Science of Peace may be possible at this time depends as much upon a definition of *science* as upon the nature of the results obtained in such a discipline. A Science of Peace is here posited as both possible and desirable.

It will be the task of the Science of Peace to demonstrate that it is through peaceful rather than warlike means that men can most easily and certainly attain their ends. This condition seems more likely to be true today than yesterday, simply because the costs of war have soared so horrendously rather than because other means have decreased in their costs. To discover that men will seek peacefully to achieve their goals we must first learn what it is that they seek. This is the first task of our science, and not the easy task which it seems at first sight to be, for when we think about society realistically we quickly note that it is filled with a confused set of means-ends relationship not easily disentangled. When we have discovered the ends which men seek we must attempt to examine the means through which they seek them and the way in which these means affect and are affected by changing concepts in science and technology, by geographic and demographic factors, and by the character of the structure which presently exists to serve wants and needs. Only then may we presume to show men how they may peacefully seek those goals which do not involve war.

To succeed, this task will require the work of no small number of men. It costs time, money, and brains. It will involve devotion to truth even while examining the most sacred tenets of civilization. Such an effort will require support from many men who are not themselves satisfied with the pursuit of truth for its own sake but who must be shown how science can make peace more probable if they are to give their support. Hence one of the first criteria for research in the Science of Peace is the answer to the question: "Does it promise results which will help to establish the value of scientific research in this field?"

It may be that the kinds of research that can gain the necessary support are so crudely empirical and so limited in their implications that nothing significant can come from them. But by concentrating scarce resources on those projects likely to demonstrate their usefulness we may make it possible later to examine more basic concepts.

It may be that despite great efforts no result sufficiently reliable to justify further support will emerge. But it seems that a Science of Peace is more likely to survive if it can help to solve present problems. To guarantee this survival another criterion for the initial effort at research might be the answer to the question: "Does it deal with immediate and pressing situations threatening war?" Since it is war which we seek to prevent we will pragmatically justify support this way. Moreover the character of the present situation is such that we must maintain peace in order to discover the prospects for future peace. The premises from which we reason today would be greatly altered as the consequence of another World War. The maintenance of peace then is required to permit examination of the probable consequences of present conditions, which must serve as the empirical basis of our judgments about the future.

Areas of immediate threat can also be used as foci at which we may discover the factors involved in the maintenance of peace and from which we may trace them outward to the conditions that beget them. Tentative hypotheses can also be tested against emergent events and modified accordingly. Such situations can serve the Science of Peace as the clinical record serves the Science of Medicine, but we must not forget that here as in clinical medicine we must exhaust all the other possible means of verification before putting our theories to the test of action.

The Science of Peace must, like all science, involve some assumptions. Among those we feel to be necessary are the following:

> 1.   Human choices are factors influencing future events. Unless this be true no science can justify itself by providing the means to gain or to avoid what would otherwise not have happened.
> 2.   Human behavior is a product of necessary antecedents. Science is possible only in stating the probability that events will recur. If there are no necessary antecedents to human behavior recurrence will be so rare as to make science useless.
> 3.   Human values, in so far as they are conceived to be an aspect of the personality of any person living at any time are for him experiential in origin. We can have no Science of Peace if men come by the values which result in war or peace by means scientifically unknowable.
> 4.   A Science of Peace is possible only if there is access to so large a part of the knowledge necessary to prediction that the major factors controlling the future can come to be known to the scientist.

It is only as a consequent of his ability empirically to test his hypotheses that the scientist can increase the accuracy with which he can state the probabilities involved in the relationships of future events. While we will not know whether it is possible to get at the necessary facts until we discover what it is that we must know to test fruitful hypotheses, it is certain that if we are unable to secure the necessary facts we must either abandon the effort here to use science or develop hypotheses for which facts *are* available.

Peace Defined

Peace as we define it is a situation in which governors of states limit their use of physical coercion to acts that are not likely to encounter extended, organized, and effective physical resistance. We do not assume that the Science of Peace will seek to eliminate all conflict. It may be that peace will come only when all of the "highest" ideals of man are achieved. If so, study of the Science of Peace will discover it. But it may also be true that a warless world is much more probable if men give up the idea of achieving universal values in the effort to secure peace. The conditions under which a warless world is possible will be more easily ascertained if the science is not burdened with the simultaneous task of discovering how other values not necessary to peace may be attained. We emphasize particularly that war and peace are products of the policies of men rather than the accidental product of their unintended acts. "Normally" the governors of states using physical coercion or its threat within prescribed "legitimate" limits are not met with physical resistance. Threat of war arises when states seek to exceed these limits and encounter effective resistance. Peace then depends upon the ability of governors to anticipate the character of resistance to be encountered when they act. The Science of Peace must concern itself with increasing the accuracy with which this can be done. It can contribute to peace by providing knowledge of less costly alternatives to secure the ends being sought through force. It can also demonstrate how ends themselves are likely to be altered by the choice of means to their attainment, so that means destructive of certain ends will not be used in the effort to achieve them.

Five Models of a
Peaceful World

Scientists have frequently found it fruitful to direct their research in terms of a hypothetical model. While there is some danger that such a model will become reified and interfere with the development of other, more fruitful hypotheses, we think that the results have justified the use of models. We may conveniently classify the conditions under which peace is likely into five conditions of society briefly characterised by the models described below:

> War is to be avoided by the fact that
> Model I. Men universally share values such that their means and ends are harmonious and attainable without coercion.
> Model II. War is so abhorrent to so many men that no resistance will be encountered by a state which seeks its objectives by force.
> Model III. Power to exercise physical coercion is so widely and evenly dispersed throughout man-kind that war is impossible.
> Model IV. Power to exercise physical coercion is so highly concentrated in the control of a few men who seek mutually compatible and attainable ends that prolonged effective resistance by others is impossible.

Model V. There is such a distribution of values and of power that while war is possible it is clearly manifest to all those in the position to decide whether or not to resort to war in pursuit of their ends (here-in-after called "the elite")[1] that war is in this pursuit inferior to other available means.

Assuming that we can subsume all possible conditions under which war may be avoided under one of these models we can undertake research to establish which is most likely to come to be. We think it possible by inspection to show the very great improbability that the first three are likely to eventuate. To test the likelihood of the other two much more extensive research will be required. We may turn first then to inspection to show which models should be used.

Model I. Men universally seek values such that their means and ends are harmonious and attainable without coercion

The attempt to establish such universal values as a basis for universal peace is a very old one. Almost all cultures which have felt the effects of defeat and subjugation in war have produced religious and moral teachers who sought to exorcise war. They preached universal values that denied the virtues of war and the goals attainable by war, while affirming values otherwise attainable. There exists to our knowledge no empirical evidence that such effort was long successful over a very wide area. Scientific research in history and archaeology have shown us no past experience in which there was a universal culture such that all men shared universally harmonious values. On the contrary, more exacting research seems to yield increasing evidence that there was in the past wide diversity in values and in the techniques for attaining these values. Nor are there many known cultures which accept as supreme, values which deny the possibility or the necessity of using physical coercion to secure or maintain other values. It is true that passivity is a virtue often preached, but those who embrace it as being final are usually occupants of a system actually dominated by others who are willing to use force if it be necessary to the attainment of their goals.

A scientist then, who sought to test this model empirically, would find little or no historic evidence he could use. Continuous reassertion of the final validity of the Christian or the Capitalist version of the coercion-free world is met by equal insistence upon the ultimate universality of the values and procedures sacred to Communism, Buddhism,

---

[1]This word is used with no connotations. It seemed impossible to find a "neutral" one, so this one was adopted despite its implications for some systems of thought.

Islam, or some other world religion. Extended effort to demonstrate through the use of historic evidence the existence of some universally accepted goal toward which all men move is met with equally impressive evidence of short-run goals or cyclical recurrence.

Moreover, if, as we here hold to be necessary, we accept values as being experiential rather than transcendental in origin then analysis and extrapolation of present trends give little ground for believing in any presently predictable utopia. Enormous diversity occasioned by modern territorial and functional division of labor, the effects upon mind of the training and discipline necessary to support and maintain science and technology, and the present existence of enormously complex systems of value offer little likelihood of the universal acceptance in the foreseeable future of some set of mutually harmonious means-ends relationships.

Only one group of scientists seems to be presently proposing a testable proposition as to the way in which such a condition could come to be. A few believers in the efficacy of the verbal symbol as a complete basis for the creation of values seem committed to the idea that with modern means of mass communication they could create one set of values over all the world at one time if only *they* had control of all the media of mass communication. Two of the postulates upon which theory rests might be tested empirically. The first postulate is that "they" (scientists?) might have control over all the means of mass communication at the same time. Assuming that control over mass media *is* a significant element in the power struggle, where is the evidence that any presently existing elite or power group could get control over all the means of mass communication of the whole world? Is it likely that they could gain control by any other means than the resort to that physical coercion for which this control over mass media is supposed to be a substitute? We think that the empirical evidence to support this postulate is as lacking as is that of the conclusion that required it.

The second postulate is that having control over such mass media, any necessary set of values might be created. While Hitler may not have had as complete control over mass media as some future dictator might possibly have, study of the consequences of the attempt of the Nazis to modify certain aspects of historically identifiable "German" characteristics through mass communication seems at the moment to give little confirmation to the idea of universal success of mass media in dominating all other forms of experience in the determination of values. The same thing is borne out by examination of Russian and Japanese experiments. Further research offers considerable evidence that identical symbols are likely to have different effects on different people, and there is increasing evidence that it is not alone by symbols that values are created. These considerations seem to us to warrant the abandonment of this model for serious present investigation since there seems to be little likelihood that it will serve as a model of a war-free world.

Model II. War is so abhorrent to so many men that no resistance will ever be encountered by a state which seeks to obtain its objectives by force

This differs from Model I only in that it conceives of the existence of some deviant groups, not sharing the universal values held by the overwhelming majority. Such groups are permitted by them, without physical resistance, to do whatever they wish. What has already been said about the lack of empirical evidence that such a situation has ever prevailed holds equally well here. Added to the difficulties encountered in Model I is that of maintaining a situation in which such deviant groups existed, *none* of whom sought goals in conflict with *other deviant* groups who were unwilling to give up their goals without physical resistance. The improbabilities of this situation seem to be overwhelming!

Model III. Power to exercise physical coercion is so widely and evenly dispersed throughout mankind that war is impossible

This model is also an old one, somewhat corollary to the first in that equality of power among men was visualized as at once an outcome of and necessary to the coercion-free world. The assumption here is that if every man possess the means to his own ends, these ends themselves being limited to those which he is personally capable of achieving, then no man is beholden to and hence subject to the power of any other. As in Model I research reveals no world in which such a condition has existed for long. Social organization to achieve ends impossible to single individuals brings into being aggregations of power such that no man alone can effectively resist that power. The achievement of the means to values widely held in modern times in the urban and industrial areas of the world of necessity requires the creation of power systems capable of rendering whole areas dependent upon them. Individuals in that area become totally incapable of effectively resisting. Short of the destruction of the industrial-urban world, which apparently would come about in the near future only as a consequence of war which the Science of Peace is intended to forestall, there seems to be no likelihood that this model could be set up.

Model IV. Power to exercise physical coercion is so highly concentrated in the control of a few men who seek mutually compatible and attainable ends that prolonged effective resistance by others is impossible, and hence war is impossible

This model has the virtue, for empirical research, that it has been approximated over various parts of the world for considerable lengths of time. As contrasted with Model I it assumes no universal acceptance of a set of internalized controls in the shape of values by which men, in search of their own self-fulfillment, are guided to act in mutually harmonious manner. It assumes rather that power to coerce can be concentrated in the hands of one elite so that by careful calculation of the areas in which the elite will seek in pursuit of their values to exercise power, they can be sure that they will encounter no effective physical resistance. We take it as established that no one elite is presently so situated. Consequently, calculation of the possibility that this model can come into effective operation involves a consideration of the conditions necessary to bring it about and the resistance likely to be encountered in bringing it about. We think that at present the emergence without resort to war of a world dominated by one group of men is, to a degree beyond calculation, improbable. But to establish the degree of probability that this model *can* be realized we must resort to empirical research. Such research must examine all the probable conditions for peace. In discovering the probability that Model IV will be realized we may also establish that probability for Model V.

Model V. There is such a distribution of values and of power that while war is possible it is clearly manifest to all those able to decide whether or not to resort to war in pursuit of their ends ("the elite") that war is in this pursuit inferior to other available means.

There has never been a universal situation of this kind, but very large areas of the world have operated under it, and are presently operating under it. This situation is characteristic of great aggregations of states such as the United States of America, the British Empire and Commonwealth and the Union of Soviet Socialist Republics. It is in fact probable that *all* the conditions it posits actually exist in the world *now* save only the condition it posits that it *"is clearly manifest"* to all elites that war is inferior to other means to the attainment of their values.

It would appear from inspection that this Model is the most likely to eventuate and that less change would be required to bring it about than the achievement of any other. To secure its operation then would require the resort to less speculation about the conditions under which it might be realized than is true of any other Model. We will therefore use it.

The use of the model requires for its demonstration a series of propositions. We must examine in each area of the world sufficiently powerful to make war the answers to the following questions: (1) What are the dominant values? (2) What is the structure by which it is decided which values shall be given priority and how much effort will be expended in obtaining them? (3) What means to these ends are available to those who determine which course of action will be taken?

It is possible that such research will demonstrate that certain widespread values are such that they will be sought, even at the cost of war, that they are compatible with or demand war for their fulfillment (such as the glorification of military behavior) and can probably be more cheaply achieved through war than through other available means. In this case our model is incapable of being realized.

It may be, however, that certain values will be abandoned or sought by other means than war once it be clear that they can be thus secured only at the sacrifice of other more highly valued objectives. It may be that within well-defined geographic limits certain values are attainable without resort to war, but beyond these limits are only doubtfully to be secured or are clearly impossible of attainment even with war. In this case abandonment of efforts to secure universal values in favor of patterns of regional dominance may be the more probable outcome.

It is possible too that research will establish that one elite exists which is so powerful as to make Model IV more likely than Model V.

The Science of Peace
Deals with Politics

What we are dealing with here is Political Science in its most exalted sense, i.e. the location of power in society and the determination of the objectives of and the limits set upon those in a position to wield power.

Within the state, as in interstate relations, there is the constant necessity of determining the legitimate sphere of state action. It is possible that by analysis similar to that which is used in domestic politics we may find means to predict in interstate relations how a state may determine the effective limit of peaceful action.

One key to the discovery of whether or not a state will be likely, in extending its power, to meet resistance will be found in the examination of the means by which the actions of states become legitimate. It is easy to demonstrate that men find themselves in situations in which part of their activities are controlled by other men, acting in the name of the state, who are permitted without resistance to use physical coercion if necessary to achieve certain purposes. Some theorists postulate that men willingly subordinate themselves to such power in order to achieve goals not possible unless all men be limited in their use of physical coercion. Other theorists hold that men, finding themselves unable to resist the force of the state, come passively to accept and to rationalize such use of force. Certainly most men find the power of the state to be irresistible, but there are areas in which they do not willingly subordinate themselves to the power of the state. Thus the problem of "legitimate" coercion remains fundamental to political science. It will be a major concern of the Science of Peace to discover (1) which actions are likely to be undertaken by various specific states (2) in which areas and realms these acts are to be considered legitimate and hence not be resisted; (3) those likely to be considered irresistible either because resistance is felt to be futile or because the use of force to resist force is denied by the values of him against whom it be used and, in contrast, (4) those acts of the state in question which are in the area concerned likely to be resisted by resort to arms if necessary.

Some take the position that emergence of interstate organization adequate to achieve men's goals without resort to war will come only where there is widespread prior adherence to a set of ideals or principles which make war immoral, illegal, and hence impossible. They would confine the use of force to only one legitimate power acting in the name of those ideas or ideals and thus bring about Model IV. To achieve peace based upon this hypothesis it will be necessary for men first to agree upon some desired set of principles, statement of rights, world constitution, international law, or binding religion, and only after that to establish government outlawing war in their name. This point of view is widespread and seems to dominate much of the work being done by many of those favoring peace. The position taken is that "Internationalism" as an ideology will serve to justify whatever surrender of "national sovereignty," property rights, religious or other freedom may be required in abandoning war as an instrument to achieve or to protect these freedoms. Acting on these postulates the scientist who sought to *discover* the *prospects* for peace would do so in terms of evidence of growing and effective acceptance of this ideology. This in contrast with the power of other ideologies such as nationalism, capitalism, and individualism which permit violent resistance of international authority when it invades spheres by them "legitimately" reserved to national states, to property owners, or to individuals. He would also examine the means by which such an ideology might be advanced and become dominant. The statesman who sought to *create* the conditions of peace would do so largely by manipulating symbols and experiences in a manner dictated by these discoveries so as to induce acceptance of the ideology of internationalism. Only then would there be an attempt to extend inter-state powers, and the necessary extent of those powers would be ideo-logically determined.

An alternative theory of the way in which the use of force is legiti-mized treats this process as a part of evolving societal organization. This theory holds that men with one set of purposes or values, themselves not always well or completely understood, create organization in the effort to achieve or protect these specific values. A great deal of such effort is *ad hoc* in character, and organization is achieved through regen-eration; that is to say, part of the gains made through use of such *ad hoc* arrangements is fed back to strengthen that arrangement, so that it may more certainly be repeated and be more effective in achieving the desired results. Thus function begets structure which promotes function which strengthens structure. A part of the gain is also used to create and propagate a rationale defending the structure. It is thus that means-ends relationships become so significant that their preservation is felt to be necessary to the group and coercion is permitted to secure that preserva-tion. It is thus that state action becomes legitimate.

However a great deal of what goes on is never completely understood, recorded, conceptualized, or described. But newcomers to the situation who must assume roles and carry out functions necessary to the preser-vation of organization must be told what *is* necessary to the preserva-tion of that structure. Ideology serves that end. It comes into being only *after* the organization has developed fairly stable structure whose "rai-son d'être" is only then, and then only partially, discernible. Ideology establishes in the minds of the newcomers in place of flexible arrange-ments reified symbols. Required change is made more difficult because of adherence to this set of reified symbols in preference to patterns more efficient in achieving the desired goals. Yet since real events require those who act to deal with them somehow or other conceptions, modifi-cations, new rationalizations and other verbal variables are frequently introduced to permit effective action by those in power.

Hence in a period of change ideology ceases to be a good guide to conduct. Similarly, it is a poor guide to prediction of the behavior of those who are supposed to follow it. This is not to say that ideas and ideology play no part in influencing values, but to insist that there are also other influencing factors. The researcher attempting to discover the roots of action is required to examine real choices in real situations to discover to what degree ideology as contrasted to other elements enters regularly in directing action. He thus discovers the degree to which ideology is a significant factor in determining those acts the recurrences of which he is attempting to predict.

It is possible, as we have already indicated, that research will demon-

strate that ideology *is* primary in determining inter-state action. Several
of the ideologies which characterize large numbers of people in various
parts of the world today each claim to represent universal supreme
values. If ideology alone be basic to organization and states can legitimately claim the right to exercise coercion only within ideologically
determined limits, then states in the effort to exercise what is by one
ideology required and made legitimate must clash with other states
equally legitimately pursuing another course of action: war must be the
necessary outcome of such a situation. The hope that peace may be
secured by international organization stemming from ideology would
seem to be poorly founded. There seems to be little likelihood that those
seeking to promote the ideology of internationalism can presently successfully compete with the older established ideologies which will continue to be promoted by great states and other powerful institutions.
Hence if ideology be sufficient cause and explanation of war there can
be no Science of Peace because peace itself is impossible and hence
scientific knowledge about its likelihood would in no way influence its
prospects.

The Science of Peace
Deals with Ideology

It becomes relevant to determine what is the actual role of ideology in
interstate relations. There presently exists adequate empirical evidence
for this purpose. The first and most obvious area which can be used to
establish the significance of ideology is in the historic facts of di-
plomacy. If the conduct of those directing affairs has in the past closely
followed an ideological pattern, if there has been consistent effort to
correct it when it is seen to deviate from that line (rather than to change
the line), if increased propaganda inducing men to accept the line has
been followed by performance more and more predictable in terms of
ideology, and, most particularly, if there is evidence that the course of
events has been more closely adherent to the pattern of ideology than
that which would have been predicted through the use of some other
explanation, the ideological thesis is substantiated. If however some
other explanation such as that of economic interest, regional or national
politics, demographic or technological influence or an eclectic and op-
portunistic course involving some or all of them seems to have been
followed, ideology must be given a less significant place as a determi-
nant of policy.

Further evidence exists in other places of the value of ideological indoctrination as a means to the creation of values thought to be necessary to international control of the use of force. Study of the activities of the church, particularly the Christian Church, should serve to clarify the issue. Missionaries have been active, apparently very largely for the purpose of creating ideology, in many parts of the world. Empirical study of the degree to which the official ideology which they were set up to propagate has continued to characterize their activities, as contrasted with the degree to which their ideological line has been modified by contact with other influences is to be found both in church history and in contemporary research into the activities of the churchmen and the laity whom they set out to instruct. For example, some present evidence indicates that the products of Christian Missionary Colleges in the Near East have adhered to Nationalism, Capitalism and Democracy. At least they teach slogans directed to that end. On the other hand, some of the Chinese alumni and some of the Christian missionaries to China themselves seem to have been instrumental in supporting the present Communist regime. A study of the actual performance of the alumni of missionary colleges may help to show the degree to which ideological orientation is effective in producing the goals at which it was aimed. Other types of research establishing the actual significance of ideology in interstate relations in the past will readily come to mind. These will *also* evidence the degree to which many of the proposed efforts to secure peace are likely to be successful.

Empirical research designed to test the significance of ideology as a means to prediction of interstate relations must attempt to account for the behavior observed. If ideology be not adequate to this end, then some other necessary set of relations must be inferred and tested. It was once widely held, and it is still thought in some places, that the most accurate description of human behavior is to be found in the concepts used by the physical and biological sciences. It is now pretty generally accepted that "Human Nature" as so described permits a tremendously greater range of behavior than is actually observable in any society. Concepts which permit more accurate prediction of this narrower range of observed behavior thus appear to be more fruitful than reliance upon those which permit prediction of only the wider limits.

A number of hypotheses about "social forces" and similar constructs drawn from physical sciences, as well as a number of theories concerning the inherited residue of experience of succeeding generations in the form of the "racial sub-conscious," the "folk soul," etc., have also been adduced and tested in the effort more closely to predict comparative behavior as between the inhabitants of various states. They seem less fruitful today than when they were proposed. Currently another set of assumptions seems to offer promise and it is upon these that it is proposed that we base the Science of Peace. This approach assumes that it is possible by studying the factors affecting human choice to antici- pate choice, which is in turn considered to be a factor influencing what is to be. We may use this scheme then to test the degree to which ideology or other factors in various areas or spheres of activity are effective in determining choice. The factors affecting choice by the individual we will term *values*.

The Science of Peace
Deals with Values

If factors other than ideology affect men's choices as to whether in a given situation they will resist or acquiesce when confronted with the effort by an elite to extend the range of state action beyond what has hitherto been legitimate, the Science of Peace must deal with all the values involved in such decisions. It is posited here that there is no special means by which values operating in interstate relations origi- nate, and no special process by which they are created, they exist in a continuum with other values. Criteria to establish the order of relevance to the Science of Peace of various divisions of the sciences that deal with value will be found in part to lie in the terms previously stated: Which will yield results quickest, which deal with those situations now threat- ening peace, and, for which are the necessary facts available? Other criteria to establish relevance will appear as the discussion proceeds.

One of the most difficult problems for any science of value is that of determining the model to be used. Characteristically in an individualis- tic society men assume that because values are attributes of individuals the use of the individual as a focus to guide the attention of the scientist will always be the most efficient model. This cannot be accepted as a dogma, for other models have been effectively used, and the question remains a moot one. One of the most frequent sources of confusion is the failure to distinguish between the act of evaluation on the part of an individual, and the observable pattern of action which demonstrates the position of a value in a pattern or hierarchy of values. We mean by the act of evaluation a "sorting out and ordering" process which takes place when events are mediated by way of the human body, particularly the cortex, resulting in preferences for various courses of action. Except through introspection such acts can be known to an observer only as they result in overt action (including of course the verbal statement of preference). Hence for a *science* of value which must deal with "public" facts it is necessary to *infer* the nature of this act of evaluation in others. When individuals confronted with a situation in which alternative courses of action exist take similar action, it is inferred that they share similar values. Stated another way, we call values those factors which, within the physical and biological limits present, affect choice. The accuracy with which acts follow a course which would have been predictable from pursuit of the inferred values determines the usefulness for the Science of Peace of the imputation of value.

When individuals make the same choices in the same situation we are warranted in inferring that at that moment they share the same values. We are not able to infer from these choices the moment when evaluation (the sorting-out process) took place. It may have been immediately precedent to the act, or it may represent a pattern previously adopted in the society, learned by the individual, and presently being carried out without calculation on his part of the consequences of, or consideration of alternatives to, the act. It is possible to set up experiments in which those choices directly consequent to evaluation can be distinguished from those which follow a previously established pattern. For some branches of the science of value this distinction may be of the greatest significance. For other purposes, where only the fact of regularity of response is significant, the pattern of choice may be determined, using whatever model most adequately serves to predict that pattern. Thus evaluation is itself not a group process or cultural process. We may find it possible, however, more accurately to predict values prevalent in a group or emergent from a typical situation through the study of patterns of recurrent behavior in groups or situations than by attempting to study all the idiosyncratic reactions to be found among the individuals who make up a group or institution, only the dominant and repeated pattern of values of which we are interested in predicting.

To say that values are attributes of individuals is not the same thing as saying that the pattern of values found to be prevalent in a society has an exact counterpart in any individual in it. That pattern is rather a result of the interaction of many individuals, each pursuing his own hierarchy, derived from his own experience, sacrificing something of lesser value for something of greater value but finding that in so doing he frequently is able to achieve his values most advantageously in some order other than the one he would ideally achieve. This arises from the fact that his culture seldom supplies him with techniques and materials at costs exactly commensurate with his wishes, nor does the hierarchy of others correspond in a completely complementary way with his own. Hence the pattern of choices exhibited by the individual may not correspond with his "subjective" hierarchy. It represents rather an adjustment to the fact that the services of others, whose values are different from his, frequently provide the means by which he can attain his own ends, and hence the pattern of choice is a consequence of interaction between him and those necessary to the attainment of his ends. Thus for some purposes, such as therapy for example, the particular hierarchy of values "subjectively" considered may serve as the necessary model. For other purposes that hierarchy may not be so good a model as the study of the pattern resultant from interaction.

It is particularly necessary here to recognize the difference between learned patterns of behavior, transmitted as a pattern together with other elements of culture, and values which represent the products of personal evaluation. In childhood learning in the family, neighborhood, and local community, values are in large part learned symbolically, in anticipation of experience with the events which they symbolize. That is to say, that value as learned in a given situation is given a position subordinate to some, and superordinate to some other, values. In an operational sense this order or pattern constitutes value. What can be anticipated in the behavior of others from the inference of value is the order in which values will be made subordinate, indifferent, or superordinate one to another. Members of families and communities are taught as children to put "first things first." A pattern of values reinforcing one another is created by the authority of those who are in a position to make "good" things "good" in terms of their immediate consequences in fulfilling the basic biological and social needs of the child. The values of the individual who is the product of such patterns become predictable and he is inferred to have assigned, in some definite order, values to acts in situations. For research into values which are largely the product of this kind of situation, the use of the individual as model is enlightening and perhaps adequate since a sample of the social product will serve to predict the whole.

Where the individual enters into more groups and situations, and these become disparate so that the unified pattern is overlaid with experiences in which the consequences of putting what were "first" things first, varies among the individuals making up the group, the meaning attached by the individual to these early experiences changes. For those who follow this pattern and get the consequences they were taught to expect, the pattern is reinforced both in overt behavior and, presumably, in psychic structure. For those who follow the pattern only to experience quite other than the anticipated consequence the result must be alteration either in overt behavior, psychic structure, or both.

No universal pattern of values is taught to children throughout the world. Since the consequences to any person of following any pattern in interaction with others seeking other hierarchies must be to a degree disruptive of the existent configuration, the attempt to predict the emergent societal pattern of the world from the study of individuals would seem to depend upon the adequacy of the sample as representative of the universe. The assumption that this model is an adequate guide both for research and for action is widely held. It is frequently assumed that changes in the character of the individual must be precedent to the changes in the world community, which is in turn held to be necessary to the creation of a World State which will limit the interstate activities of groups to those felt to be legitimate. Upon that assumption a great deal of effort has been expended, not only in support of research but particularly in actions designed to induce individuals to change their values.

For the Science of Peace this poses the two-fold problem of determining

1. the actual usefulness of the model in research;
2. the probable outcome of action based upon it.

It is obvious that research demonstrating one of these propositions relates to the other.

It is least questionable whether or not the basic assumptions upon which the use of this model rests are as adequate as some others for the prediction of emergent patterns of value in secular, urban, and industrial society, however applicable they may have been in dealing with social processes in smaller areas involving fewer role relationships and less secularized controls.

Associations and institutions such as the national state and "functional groups," by which we mean groups devoted to a limited number of fairly closely defined objectives, such as trade unions, merchant associations, insurance societies, or corporations, constitute value-creating, value-mediating structures which differ in very significant ways from the family or local community. The number of such differences is large and has received considerable attention. Of great import here is the fact that such groups attempt to impose no complete pattern of values as the "primary groups" do. That is to say, they tend to neglect or disregard many of the values of the individuals which make up the group. They emphasize the primacy of the values which *they* exist to serve. In Western states no institution other than the family exists to order and give expression to the needs of the "whole man." "Political," "economic," "religious," "recreational," "aesthetic," and other values are promoted by institutions deliberately separated one from another in terms of their functions, and that separation is rationalized as a part of the moral, religious, and legal codes. Consequently, the *pattern* of values *discernible* in such areas will be resultant of the claims at that time of all such groups and institutions, operating at that point, in addition to those patterns previously imposed earlier upon the individuals involved by the family and local community.

The functional organizations seek only to insure that the value hierarchy of their members shall include in a favored position the particular values which these functional groups are trying to achieve. It is of no concern to them that the hierarchy characteristic of the individuals who make up their membership varies considerably. Because of this disparity among the members the pattern that emerges as that being sought by the group may vary immensely from that of many of its members. When in turn this "unrepresentative" hierarchy interacts with that of other groups of which an individual is also a member, the disparity between the pattern of value of any one member and that which results from interaction of groups of which he is a member may be enormous.

Thus the individual may continue to hold his "own" hierarchy intact as an "ideal" and give lip service to it in the expression of opinion. He may at the same time find that the order in which he can in fact realize his values at the least sacrifice of other values is quite different from that which he would achieve if he were free to act without the use of any functional group. Thus the system of "learned values" which constituted his childhood pattern and may be considered "basic" by the depth psychologist may be replaced by acts of evaluation which result in action which considerably differs from that pattern. Psychologists may then differ as to which constitutes his "real self," but for the Science of Peace inference of value from the recurrent pattern of observable action may offer more adequate basis for prediction than the study of the ideal hierarchy of the individual.

What we have said may be restated thus: because the available *means* to action has a great deal to do with the actual choice of courses of action, it is very difficult to predict action from individual hierarchy. Models using the individual process of evaluation for predication of resultant choice are extremely complex and seem to involve for their demonstration a kind of mathematics as yet undeveloped. In general, they proceed from the construction of some scale of possible choices on which the point is taken as that representing equality of value between goods. Ends are said to be equal in value when the individual is indifferent as to which of them he shall attempt to satisfy. In other words, the means to their achievement are equally costly in terms of the sacrifice of other values required for their attainment. From a static, subjective point of view derived from introspection, each putative course of action "has its own value," and shifts in value come only as the subject is induced to change that subjective evaluation. Relative to position in an inferred hierarchy, which is what the scientist must work with, a change in value may come about in this way or from an alteration in the position of values which must be sacrificed to obtain it (since it is only in terms of its relative costs that we have determined its position). Thus end values may "subjectively" be held to be equal in terms of their costs but the costs while *equal* may be *different.* A change in the cost of obtaining one value may thus alter its position while "subjectively" speaking the value remains the same. If the attainment of one of these "equal" end values can now be secured at less sacrifice of other values than was previously the case, it will be chosen over another even though "subjectively" there has been "no change in value." Now a very great number of changes in the modern world have altered the traditional *means* of obtaining many of man's values. Some of the "sacred" values of another time have by the development of science become secularized, that is, they are considered as alternatives to other values, and also the means to their achievement can be rationally considered in terms of functionality to achieve them rather than as ends in themselves. Some values previously secular in this sense have become "sacred"; they cannot be considered merely as alternative means but are part of a pattern which will be preserved intact even though an otherwise less costly means offers itself. Moreover, there has been tremendous alteration in the costs of means, secularly considered. Consequently, to predict the order in which goods will be chosen requires an examination of all of the values for which those whose costs have changed were equivalent, or to establish new configurations of value of which they were a part. Also, most significantly, changes in technology vary in their capacity to create or mediate various types of value. For some values modern technology offers new and less costly means of attainment; for others it offers no "cheaper" way to secure them. For some it offers new methods of creating anticipatory responses; for others it is no substitute. In part, this reflects the values of those who invented technology to serve their purposes; in part, it is the unforeseen consequences of technics, not inferable from the values of the inventory. (We will later discuss more adequately the role of technology in this respect.)

The difficulty of making an estimate of such a complex system of interactions in terms of the individuals involved has become particularly apparent in the study of economics. Efforts to explain actual economic behavior through study of individual value hierarchies have largely been abandoned by the institutional economists. Value is by them inferred from the pattern of behavior exhibited when individuals and groups interact. In this endeavor it has of course become apparent that there is no pattern of specifically "economic" values. The inference of the pursuit of such values is now more, now less, useful as a means to predict behavior than is that of some other values. It is here held that this will also be found to be true in predicting the behavior that is involved in peace or war.

Opinion polls and attitude studies that permit the individual to express choice without at the same time confronting him with the probable sacrifices involved in implementing that choice in a real world may be a poor guide to his activities in situations in which such costs become apparent.

It would appear then that a model in which realistic consequences of choices are made known would be a better guide to prediction than one primarily representative of an "ideal" hierarchy. Groups would thus serve as models from which we could infer the values significant for the Science of Peace.

The use of the group as model does not easily yield more accurate prediction of values emergent from situations than does that of the individual. Initial effort in the use of group analysis quickly reveals the difficulty of discovering the ends there sought by various members. We must recall that the inference of value is for the purpose of more closely predicting future behavior. If all that we learn by inferring value from pattern is that under the given conditions pattern will be repeated we will not be able to predict how, when the situation changes, various members of the group will react to that change. We will need to discover in the group the order in which various members seek to achieve values by the order in which they appear willing to sacrifice other values to obtain them. A trade union, for example, may serve only as a means to high wages and better working conditions for many of the members. For others the union may constitute a way of life and be an end in itself. For some it may serve recreational or fellowship value. It may serve the bureaucracy of the union as a means to other ends than those of the membership. To certain factions it may be the means to the creation of the Communist world. Interaction within the groups will determine which of the values of group factions will be higher, which lower, as an objective in the pursuit of which the power of the whole association will be used. What must always be realized is that the choice to use *group power* for "higher" ends commits that power so that it cannot subsequently be used to achieve the lesser ends of the group. Similarly those who gain control of the corporation or of the national state not only are able to use the power of those who agree with them but also may control immense power derived from the acts and the possessions of those who are opposed to them and to many of their values. This situation is to a degree present in all groups. Consequently there may be in group action *no* observable resultant of the fact that some of these subordinate values are held by members who belong to it. The economist has not discovered a way to make significant in the market the order of choice among goods which are never bought, nor the political scientist to make meaningful the choices of the voter among the candidates or issues for or against which he did not vote. Hence even when families, sects or small groups continuously reproduce among their adherents patterns of value different from those which result from the interplay of these values with those which are dominant, there is no evidence that these minor themes will have any effect in changing the dominant values of the groups to which they belong. Some of these presently subordinate values *may* become bases for group action, as when an even balance among proponents of other values permits a minority to insist upon the implementation of its own dominant values. Some such values *may* become in time more significant to those who hold them than alternative values which are presently dominant in determining choice. Techniques for distinguishing such emergent values from those destined always to be subordinate are highly relevant to the Science of Peace. Pending the ability to discover which are which, there will probably continue to be considerable effort expended in an attempt to change the position of values permanently destined to remain below the level of choice for action by any significant group.

Groups in action must choose in terms of the dominant values sought and the *alternatives available to them;* group choices then reflect not only the end values sought but also the power which the group as a whole wields to secure its ends. The pattern of action represents the equilibrium of the consequents of value-directed power shifting to meet power directed by others to the achievement of other values. In stable societies the pattern becomes predictable as a consequence of the fact that, the sources of power and influence having remained relatively fixed, the equilibrium reflecting the past effort of various groups to achieve values and to induce others to accept them is also fixed. Rationalization and justification of this equilibrium becomes part of the culture, and the pattern as a whole can be handed on to the children, who identify themselves with the taught pattern. In a changing secular-scientific-urban-industrial society the pattern represents no fixed hierarchy easily discoverable in institutional stability. It remains as fluid as the various forces which give rise to it. The pattern as a whole exists in no one's knowledge. Only a small part of it can be symbolized and taught effectively either to children or to adults since it consists in large measure of situational judgments continuously made. Hence the prediction of patterns of value is a study in dynamics. It requires us to discover at once the ends pursued, the structure presently available to serve those ends, and the power wielded by various groups, each active in the effort to change the pattern so that it best serves those values presently dominant in the hierarchy of those who control the resources of the contending groups.

The most effective model for research useful for the Science of Peace will be that which reveals in the specific *area under examination* the character of the pattern of values presently extant. In some stable area the use of the older research models may suffice, but for other areas a new type of research model may be more fruitful or less costly.

There may be in every major region of the earth some basic patterns of value which are being carried along from generation to generation by the "automatic" process through which the habits, attitudes, values and other elements of the personalities of parents are transmitted to children, reinforced by interaction with other children of similar parents and all of the other experiences in a local situation which become part of the process by which the first "self" emerges. Such patterns are likely to be revealed by studies in "Basic Personality Structure." These may be extremely significant for understanding of patterns of value recognized as "national character" or "religion" or "class." In local affairs and within the territory occupied by the "nation" or amongst the groups sharing a "religion" or those who make up a "class," these values may not be apparent to those who share them. They are as "natural" as the speech by which they are mediated. They will become apparent only when some change demanding that they be sacrificed to some other value brings them to the awareness of those sharing them. Frequently even then they come to consciousness only in a sense of anxiety or undefined emotion. Channelization of this emotion into specific courses of action will usually be undertaken by various groups seeking to identify the values which they pursue with "the good of the nation" or "basic principles of the religion," or the "rights" of a "class." Hence studies of basic personality structure by "outsiders" may more accurately reveal the basic value structure of a class, nationality, or religion than is possible by those who actually share it. Prediction that action following disturbance of this structure will take lines rationally calculated to preserve these values is, however, another thing. In the case of most of the elements contributing to the "Basic Personality Structure" there is little conscious awareness of the means by which the experiences contributing to it are created. Nor is it known how best they may be preserved. These means are entangled with ceremonies, techniques, habits, and other elements of experience and emerge as a "natural" product thereof. Hence when they are disturbed there is frequently no agreement as to the means to re-establish them. Moreover the elements that make up this basic structure vary among members of the society, regions, and probably among socio-economic groups or classes.

Even to predict what for each group in a nation the elements of the "Basic Personality Structure" are would require very extensive sampling, and might show that such basic elements as could be translated into action would vary considerably among those sharing it. What one group or faction conceives to be the very heart of religion, national character, or the rights of a class may, by another faction or neighboring group, be considered of lesser significance or even abhorrent to those basic values.

If an elite be contemplating action it may be of the greatest concern that the action proposed violates some of these deep-seated values. If the Science of Peace is to provide information on which elites can reliably act, it must be prepared to reveal to them whether or not proposed acts are likely to arouse strong emotional reactions. For this purpose study of basic personality structure may be invaluable. On the other hand, it may be just as necessary for an elite to know what course of action is likely to follow from the emotional disturbance of the population or group in question. For this purpose study of groups and associations may better serve. What specific groups in an area are likely to do when aroused may then be better predicted from the study of group structure than from the study of personality structure. Choice of model will then depend on the area under observation and the character of the issues presently putatively leading to war.

Complete knowledge of all the factors affecting patterns of values would probably be too encyclopedic to be secured for all possible situations. It would certainly be too cumbersome to serve as a guide to those whose action must be taken on the basis of their present knowledge. To be useful then we must know the groups significantly involved in the situation under examination. To guide the elite the Science of Peace must be able to answer the question: "Which values of what groups determine for each of the agencies presently involved in this situation the action it will attempt to take?" The agencies involved in interstate relations will include not only international bodies such as the United Nations and national states but also cartels, international public unions and such bodies as the Cominform, the international Church, international "news" agencies, and the film industry.

Even within a single state it is difficult to discover the groups who influence specific policy. There is frequently great effort to hide evidence of such influence as is locally regarded as "illegitimate." Political Scientists, using the state as focus for their attention have, however, demonstrated that the study of such influence leads to far more accurate prediction than is secured by looking for the answer in terms of broad generalizations about the general will, general welfare, and national character. It is a commonplace that certain policies of states are largely influenced by trade unions, others by business men, bureaucrats, militarists, industrialists, professional politicians, farmers, clergy, etc. As we have indicated, these groups may also share common values not visible in domestic policies which become so, however, when their sacrifice is attempted in the interest of achieving the aims of other functional groups. Matters of domestic policy then are constantly competing in the same arena with those involved in interstate relations.

The use of the group as model for prediction may well serve to show how the power of a group will be used. Such a model may not be the best means of determining the consequence of the use of that power when the pattern exhibited is in fact a function also of the aims and the power of other groups, seeking other objectives in that particular situation. Here again it becomes apparent how the means available, including power, influence the pattern of choice. To predict which values will be served we must study what goes on where the interaction takes place.

The Science of Peace
Deals with Social Structure

The position of various groups and individuals at the point of their interaction is of the greatest significance for prediction of the resultant action. If there exists no structure to formulate and give expression to the half-formed vaguely sought values of the multitude, while the defined and articulate goals of a minority are easily given expression and its power is so organized as to be effective, the emergent pattern of action will reflect much more certainly the latter than the former. Systems of representation are designed not only to give expression to groups but to weight the influence of these groups in decision making. Codes specifying the place and manner of determining policy, structures designed to clog the flow of some kinds of information and to give adequate expression to others, rules for control over meetings, and methods of communication designed to create or impede effective group action are part of the power structure of every society. Administration of policy also provides opportunity to modify even more the "pure intent" of the member of society. These structures differ widely in their purpose and in their effectiveness in giving control to one group or another.

Students of international relations have made considerable progress in devising or analyzing schema for the structure of interstate action. A great many such models are divorced from an analysis of the real and potentially real values likely to be served, and from the power that exists to serve them, and hence are little likely to be realized. Others reveal clearly the realistic analysis of the values to be secured through interstate organization and are designed to give weight to values proportional to some real estimate of the way in which these values can and will be pursued or defended by various states. Similar research outlining the structure of some of the older national states is available. There is less adequate knowledge of other structures clearly significant for a Science of Peace. The character of social structure in the Soviet Union is not at all clear, at least to most Westerners. The emergent structures in Japan, Germany, China, and India particularly need careful scrutiny to reveal how values are to be there weighted and what dominant values seem likely to be served by the new structures. There is even less adequate knowledge of other structures significant in international affairs. The place of such a body as the College of Cardinals of the Roman Catholic Church, and the significance of the structure of the Church in determining its policies are important in analyzing the vital role which it plays. Corporations and international cartels have operated under a cloak of secrecy labeled "private business," but their policies are immensely significant for international peace.

Intimate and detailed knowledge of the social structure of "Hollywood" and the means through which control over the content of films is determined may show first how both the conscious-determined efforts of those in control, and the implicit or derivative consequences of pursuit of those efforts are important for peace. Research into the power structure of international news-gathering and disseminating agencies should likewise reveal the intent and the consequences of the values pursued. Such studies should be examined on a level with those of "government" itself for they frequently involve relationships which it becomes the "legitimate" obligation of states to protect even at the cost of war. The same thing must obviously be said of the agencies of the Communist International.

This survey merely illustrates the criteria for determining the relevance of social structure. Such structures as we have cited have clear relevance for the Science of Peace, for they help to predict the results of contemplated acts by elites. Other structures such as those of the family, the community, local government, small business, and fraternal organizations must, if they are to be considered relevant, be shown in each case, by clear linkage, so to be. The fact that they alter or preserve values not clearly involved in interstate action cannot *per se* be regarded as a demonstration of relevance.

Perhaps we can now state some general conclusions about the usefulness of various research models for the study of values. The total pattern of values can be inferred only from the study of the total configuration. But many insights into the way the pattern will change under varying conditions can be gained by the use of other models at the appropriate time. There can be no general acceptance or rejection of the use of the individual, the group, or the social structure as model. Rather, choice must be exercised to deal with the specific situation in which value is to be predicted.

We assume that through the use of such models research will make it possible to discover the values of the elites whose decisions it will be necessary for the Science of Peace to modify.

The further development of the Science will depend upon what those values turn out to be. Let us subsume these values into three categories:

(a) The ends sought are incompatible with war, which will be avoided at any cost in terms of the sacrifice of other values. In this case there is in the area in question no danger of war and the Science of Peace may turn its attention elsewhere.

(b) War is to the elite here found an end in itself or a necessary means to the ends of the elite in question.

(c) The ends sought by the elite are compatible with war and with other means. Decision as to which will be chosen depends upon which seems to involve the less sacrifice of the values of the elite.

If condition (b) exists, there is only a limited function for the Science of Peace. It is not presumed that the scientist himself will have the means to alter the values present. It will be necessary then, if the course of action is to be altered, to demonstrate how the elite in question may cease to be in a position to determine whether or not there is to be war.

Among the factors which an elite considering alternative methods to achieve the maximum of its values may consider is that of retention of its power position. Particularly in the case in question, it becomes obvious that the choice of war is possible only if the elite remain in power, so the real choice that arises is one between losing power and with it the means to achieve what would be sought by war, and retaining power under conditions which may involve abstention from war.

It may be possible in some cases for the Science of Peace to demonstrate that if an elite choose war it will itself lose power, as a consequence of the loss of support of some groups who share some of its values but will not seek them at the cost of war.

It may also be possible to show how conditions beyond the power of the elite to control may be altered in such manner as to cause it to lose power to make war that war becomes more costly than other means to the ends sought. Such means might include blockade or boycott and other forms of interdiction of trade. It might include withdrawal of cooperation in providing interstate services such as communication and transportation necessary to some groups supporting the elite. It might also include support of sabotage, incitement to civil resistance, and the use of mass propaganda.

Research which will determine the degree to which the present position of an elite within a state is dependent upon situations which it is possible externally to alter thus becomes relevant to the Science of Peace. But such calculation is also necessary to serve where conditions cited under (c) exist. That is, it is found that:

> The ends sought by the elite are compatible with war and also with other means. Which will be chosen depends upon which involves the less sacrifice of the values of the elite.

In this situation the function of the Science of Peace will involve a demonstration of the relative costs of the various courses of action.

The Science of Peace Deals with Means to Secure Sought Ends

So far we have treated values largely as a socio-psychological system, although we have alluded to the fact that the pattern of values must be *inferred* from repeated behavior. Such behavior involves a choice among the means available to secure ends. But behavior in an empirically demonstrable world not only involves psycho-social phenomena but also operates within the limits imposed by things. Values are a function of the means necessary to achieve them, and means to implement most values include things. Changes in relationships with things will be reflected in the consequences of acts, which in turn may result in evaluations different from what they formerly were. The prediction of the characteristics of emergent values and of social structure cannot accurately be made without examination of whether or not there is likely to be a change in the consequences of the use of things to secure valued objects. We here allude to no mystic connection between choice and things based on some "materialistic dialectic." We merely repeat the assumption that in a predictable world changes in the results of acts in terms of the values in which the acts were oriented lead to changed judgments about the desirability of altering or repeating those acts.

Hence changes in technology have the effect not only of changing means but also of putatively altering the position of ends. One resultant of changing technology is then a change in the pattern of values which emerge from experience with it. It is probable that a great many developing interstate relations are involved in the consequences of this fact. Both the major blocs now contending for power in the international arena suppose that they will be able to achieve the values they seek only by changing technology so that it will be possible to "raise the standard of living." This suggests a question both as to how present values must be altered to permit the required technological changes and also as to what the consequences of the proposed technological changes are likely to be. Since some elites seek to make technological change it will be necessary for the Science of Peace to examine the probable consequences of attempts, in various areas of the world, to change existing technology. Among the means to implement its values we here posit the use of force.

The Science of Peace
Deals with Force

Men who value freedom and abhor coercion find it difficult to conceive of force as a neutral means to war and to its alternatives. War represents the effort of a state, in the face of physical resistance, to extend by coercion the sphere of its operations. Whatever force be required to obtain an objective by force may be also available to obtain it by other means. Thus the energy costs required to secure an objective may be a neutral means to measure the alternatives available to those who are willing to consider physical coercion as an alternative means to secure their values.

Present calculation of energy costs of the attainment of values is exceptional. It has generally been assumed that monetary costs represent an adequate measure of the different courses of action, pecuniary valuation serving as a universal measure of value. As the use of force has been extended, it has become increasingly apparent that, however flexible an instrument money may be for the calculation of "exchange" value in a situation where other values are universally held constant by such institutions as property and law, it proves to be less than adequate when great institutional differences characterize the systems in interaction. To create two bombs or battleships of equal military power in two different areas may involve gross differences in costs. (We here allude to the "real costs" of producing the goods and services necessary to induce those who build military goods to do so, not to any difference which merely reflects difference in "foreign exchange.") Such costs will be reflected in the price of the weapons, but will in no way represent the degree to which they will be effective in the arbitrament of war. "Economic" costs in one area may represent "free goods" which have no effect in terms of price in another. For example, one state may provide for unemployment insurance, housing, workmen's compensation, and old-age pensions as a part of the "costs" of a tank while another will include none of these costs in its monetary accounting. The monetary costs of various courses of action in various areas of the world reflect a complex system of values in flux. Money costs frequently fail to reflect accurately the alternatives to be found within states. They present an even more distorted picture of costs as between states. This situation has become more common because of the various systems of fiscal policy, deficit financing, taxation, and subsidy payments, which characterize the economics of modern states. Calculation of the energy costs of courses of action producing the goods necessary for war also results in some distortions, but energy is probably a better measure than is money of the likelihood that those who control it will be likely to achieve their objectives when force is one of the alternative means to secure their ends.

To ascertain energy costs of courses of action involves knowledge of science, technology and geographical resources. It is also necessary to know what is required of technology by the existing values and social structure of the area, in which the calculation is to be made. There is a vast amount of reliable information as to the location of natural resources now necessary for the operation of technology. Techniques for the discovery of such additional resources when necessary are also well-established. Similarly, scientists can accurately predict the energy consequences of the use of a particular converter and now have adequate means for the discovery of new energy converters. The Science of Peace can utilize relevant knowledge from these fields when needed without grossly disturbing present concept or techniques.

There are, however, no similar bodies of relevant knowledge or efficient models for the discovery of the nature of the system of values or social structure necessary to the adoption of specific technics, nor of the resultants of such adoption in terms of the values or social structure putatively emergent from the adoption of technics.

The Science of Peace Deals with Technological Change

Speculative hypotheses and reified dogma for the most part take the place of demonstrable propositions in this field. It is generally assumed in the West that its technology will eventually become universal and will give rise to institutions similar to those accompanying it in England, the United States, and other areas whose economic and political institutions are largely British in origin. This easy assumption needs immediate and definitive examination. The Marxian derivative of this hypothesis is supported with even less empiric verification. To establish the degree to which the future course of events is likely to correspond with either the orthodox or the Marxian model for predicting the social consequences of technological change, we will examine two propositions:

(a) The conditions under which particular changes in technics can be effected.

(b) The necessary consequences of those changes.

With this knowledge we may be able accurately to predict the regions which are likely to become industrialized, and the amounts of power which they are likely as a consequence to be able to exert in pursuit of their objectives.

We may also learn something of the way in which present values are likely to be altered by these putative changes, and what changes in social structure are likely to take place so that we can anticipate how that power is likely to be used.

So far most international action seems to have been based upon the assumption that western technology will spread all over the world. One of the major blocs contending for power assumes that this must have the effect of creating a democratic and capitalist world. The other assumes that it will eventuate in those institutions and practices foreseen in the ideologies of Marx, Lenin, and Stalin. Empiric research may establish how adequate either of these models has been by utilizing "retrospective prediction." It is generally presumed by both ideologies that men are largely motivated to pursue material gain and will willingly adopt any system that offers promise of such gain.

Limited present evidence demonstrates that Western technology has not easily been transported to other areas. Its spread has been accompanied by continued resistance. The great bulk of the people of almost every region into which it has been thrust have seemed to prefer their old technology rather than to abandon their way of life for the rewards promised to them in return. The technique by which Western technology was spread seems to have been that of searching out dissident or deviant groups and individuals in an area who were willing to pay the price of dislocation and disorganization of their society in order to gain the power and/or wealth which resulted from their cooperation. Failing that, the "native" populations were destroyed or were forced to migrate. The trader, soldier, or missionary found some groups willing to cooperate and supplied them with the wherewithal to overcome the resistance locally arising in reaction to the proposed technological innovations. These groups were supplied with "outside" support so long as it was necessary, desirable, and possible to their (the Westerner's) ends. In some areas this effort resulted in such a transformation of local social structure and values that it continued to maintain itself when that support was withdrawn. In other areas the energy generated in its defense was inadequate to maintain it in the face of local opposition, and as soon as it became possible to interdict "foreign" aid, the processes of industrialization were stopped and in many cases reversed in order to permit the reappearance of older institutional arrangements.

Recent Western history is largely a record of the successful experiments in introducing industrialism, although such states as Spain, Italy, and Portugal must certainly offer testimony that failure to industrialize is even there to be encountered.

The inequitable distribution of the costs of technological change is characteristic of such change, since in so far as possible the innovators extract for themselves from the new technology all possible advantage, assessing the costs on those unable to avoid paying them. It is the business of the scientist to discover for each area where change is proposed what these costs are likely to be, how they will be assessed, on whom, and what the resultant reaction is likely to be. These facts cannot be discovered merely from the study of "economic law" or "human nature." They must be discovered for each region in terms of the values which must be sacrificed and the values likely to be gained there and the likelihood that the change can successfully be made without war.

In the West it has long been expected that groups endangered by the free spread of technology should organize to prevent that spread. Hence the origin of tariffs, patents, copyrights, trade marks, trade agreements, trade secrets, and similar instruments of the propertied as well as perhaps the less familiar devices by which priests and peasants, medical men, and mechanics have operated to protect their values. It should not be surprising then to note that often the very processes of technological change endanger values and thus generate reaction sufficient to bring into existence groups powerful enough to stop its further modification and spread.

The resort to history does not seem to establish the validity of the assumptions upon which the old models are based. To some degree any research in this area must be speculative. The nature of the process of social change is not sufficiently well established that we can know in advance of adequate research just what factors are most significant. There does exist sufficient evidence as to the conditions under which technological change will take place to make irrelevant many of the generally accepted dogma which are widely used as bases for action. The fact that science and technics have been sufficiently available in the West so that any interested group could borrow them cannot be disregarded. The rapid and extensive use of this science and technology by such groups as the Japanese, Germans, and Russians is noteworthy. Failure of other areas to make use of modern technology must be found either in terms of force from outside the state supporting values and social structure antagonistic to industrialization or in terms of local effort to preserve local values and local structure which would otherwise be modified by technological change. Critics of Western Imperialism have adduced that technological change in many areas was prevented primarily by Western intervention. While definitive evidence to the contrary does not yet exist, the fact that these same institutions of Western Imperialism were unable to prevent—and apparently were actively an aid in promoting—the spread of technology in such areas as the United States, Canada, Japan, New Zealand, and Australia indicates that if imperialist policies were necessary conditions for the failure of technological change in much of the colonial world, they were not sufficient explanation thereof. It is thus apparent that other factors were at work in many areas of the world to prevent the general acceptance of Western technology and its resultant power. The examination of these other factors is highly relevant to the Science of Peace since the balance of power and to some extent prediction of the character of values to be pursued in interstate relations depends upon knowledge as to whether unindustrialized areas are to adopt technology which equips them with great force as compared with their present military weakness or whether present values are to continue to be served at the price of continued military weakness.

Even if a definitive theory of technological change existed, it would probably not require that the character of change in areas not now industrialized would correspond with that which took place earlier in the areas now undergoing technological evolution or presently dominated by industrial technology.

A number of conditions already demonstrated to be significant for such change are now quite different from those which existed during the Industrial Revolution in England. To select a few of such variables for illustration: There is presently no great undeveloped and unoccupied arable land such as existed in 1650: The Population base in the unindustrialized areas of the world is immensely greater in proportion to their present land holdings than was true of those areas since industrialized: Then the state of hygiene, sanitation, and preventive medicine was not such as to permit the survival to old age of a great proportion of the children born: The state of technology at that time was not such as to permit immense accumulation of force in the hands of a very few persons relative to the force capable of being mustered by the multitude: The state of science and technology was not such as to permit metallurgy and synthetic chemistry to substitute for exploration and trade as a means to obtain some materials necessary to industry: Such structures as the national state, the international cartel, and the great industrial unions were not as widespread at that time, nor was the ideology of nationalism as widespread.

The relative significance of these facts and other conditions which were also then true and are no longer true must be established as the theory of social change in relation to technological change is developed. But it would be altogether unrealistic to assume that facts of this character are of little significance compared with some yet undemonstrated "inevitable" force in determining the prospects for industrialism in the as yet "undeveloped" areas of the world. We indicate something of the significance of the facts just cited both to show how they may affect future prospects for industrialization and to indicate areas of future research, relevant for the Science of Peace.

Many of the advantages of industrialization accrue largely as a consequence of the substitution of energy derived from coal, gas, petroleum, and falling water for that of plants and animals, and the use of the flowing stream, and the sailing ship as efficient means of transportation. Consequently, the efficient exploitation of these new sources of energy and of the location of the resources necessary to the new technology require a different distribution of population than was resultant from the use of the old sources. In industrialized areas fewer men are needed to produce necessary food, more men are necessary to man the machines centered in the new sources of fuel. Thus, one of the necessary steps in the process of industrialization is the transfer of population with a present claim to the product of the land to areas in which their previous claims are neither functionally retained nor traditionally sanctioned. Removal of a man from the farm or village to a distant factory site requires that the food which he previously was entitled to share because he shared in producing it, or because he was a member of a family or other locally sanctioned organization justifying his right to that share, must now be provided him on some other basis. Industrialization of the West was made by transferring to the factory site surplus food from areas in which local claims did not require its consumption there. In part, this transfer was sanctioned by the feudal claim of the landlord which permitted him to dispose of his share of the product of the land as he saw fit. The far greater part was secured where necessary by inducing part of the population to emigrate to places where such food was produced by new techniques on newly cultivated land under conditions such that the sanction of newly established social organization justified its removal. In the East today the industrializer is faced with this problem in areas in which all the presently arable land has been for centuries cultivated under a system justifying only limited removal of food from the vicinity. To industrialize he must then force a large part of the local population to migrate or to starve so that food for industrial workers located elsewhere can be provided. He cannot hope in the future to make this change even with as little resistance (which was far from inconsiderable) as confronted the British industrialist in the Eighteenth and Nineteenth Centuries.

This problem will be further accentuated by the nature of the population base and the character of its growth. When industrialization was originally undertaken, the population base in the West was comparatively small, with a correspondingly small increment of growth derived from a slight increase in the survival rate. It was thus possible to deal with this increase, as well as the necessary translocation of population, through migration. With the present enormous population base in areas such as China or India, a very slight increase in the survival rate results in an enormous increase in population. To solve this problem in the future in the same way that it was solved in the past seems to be impossible.

The fact is that the spread of Western technology to unindustrialized areas has been perhaps most easily and rapidly accomplished in the application of preventive medicine and other aspects of public hygiene. The increased rate of survival in these areas from such techniques has tremendously accelerated population growth, so that increases in food supply resulting from the application of Western methods of cultivation, irrigation, fertilization, plant selection, hybridization, and other agricultural techniques have been more than equalled by increased demand for food. This makes the problem of establishing new claims for food for industrial workers as against the growing rural population even more difficult of solution.

Factors determining the growth of population and its spread are extremely relevant knowledge required to calculate the prospects for industrialization. The factors affecting fertility are immensely significant and are as yet almost completely unknown even in the West. Compare for example the ease of learning epidemiology and other elements of "death control" with that of securing knowledge of effective means to increase birth control. The easy assumption that the problem of population will "automatically" be solved by the growth of industry as it once was thought to have been solved in Western Europe and the United States has been shaken by the course of events in Japan, Germany, and Russia, and it requires re-examination even in the United States. Nor can it be assumed that even were these areas to be able to solve their problem a similar solution could be adopted in other areas where the values and practices associated with birth control seem to be even more basically in conflict with prevalent religious and moral values. So far, in fact, it seems that only in Protestant countries of the West has limitation of population taken place early enough to avert conditions almost prohibitive of industrialization without very extensive use of "illegitimate" coercion; that is, coercion not sanctioned by generally pervasive local morals and religion.

Another factor rendering it unlikely that the same course of events will accompany further industrialization as it did in the past is to be found in the changes in values and social structures in presently industrial states as compared with those which prevailed there when industrialization first began. At that time "economics," morals, and law justified an immense transfer of the products of the energy of coal and falling water into other regions. The products of industrial England were poured out wholesale into colonial lands in return for raw materials and promises of future payment. The standard of living in England rose but not at all commensurately with the increase in productivity. In the struggle for "foreign investments" and in the absence of control over competition among exporters for "profitable markets" in "backward" areas, these areas were cheaply supplied with products resultant from the expenditure of immense amounts of energy derived from the British coal mines and water falls. Subsequently they derived other large amounts from other industrial states. Today, however, nationalism sanctions and requires the distribution of much of the product of industrial states to supply the demands of the domestic producer of raw materials. It also demands enormous amounts of energy to operate collective services such as road building and maintenance, free schooling, and the care of the indigents. Even greater demands are made to carry on military activities and to pay veterans and others for service in connection with past wars. In place of the "free competition" among capitalists for foreign markets we have on one hand the use of the cartel or marketing arrangements which permit the industrialist to maintain prices in export markets above the level which would be required in a free market, and on the other hand the close supervision of foreign investment—or refusal by nationalist governments in the former "backward" areas to permit investment under those conditions apparently necessary to secure investment. Also, there is the action of the powerful labor unions forcing the industrialist to pay out much of the increased product to be spent by workers on domestic consumption goods. Under such changed conditions it would seem unrealistic to anticipate a return to conditions similar to those which originally accompanied industrialization. What needs to be emphasized is that these arrangements were made under specific technological, scientific, geographic, demographic, ideological conditions in pursuit through specific social structure of a specific set of values. Many of these factors no longer prevail, and as a consequence we must re-examine theory to see whether or not these changes require the adoption of a new set of propositions about the relationships among science, technology, and society.

Among the changes of outstanding significance for the Science of Peace are the geo-political consequences of changing technics. A great deal of the energy developed consequent to the industrial revolution has been expended in defending the places judged to be significant for the perpetuation of the Western social system as it developed during the last five hundred years. This social structure and the geographic areas strategically necessary for its preservation were an outgrowth of the utilization of the sailing ship. Apart from this relatively powerful means of transportation, production was during this period primarily based upon the exploitation of the energy of men, plants and animals. Now plant and animal life, because of its variance with climate, is likely to differ in direct proportion to latitudinal difference between two areas. It is probable that two valued plant products will differ in scarcity in two widely separated geographic regions, and this relative scarcity will result in a large margin of profit for the trader if he can cheaply transport them. The sailing ship, using cost-free fuel, filled that need. Social arrangements were developed which permitted the elites of the trade-connected regions to maximize such of their values as were served by trade. Areas in which local values interdicted trade were frequently victimized by the increased wealth and power of their trading neighbors. What "grew up" was a system favoring values which may best be served by means of trade. These values and the arrangements necessary to their preservation became the unquestioned objectives of the trading elite, and in so far as they could be influenced by the trader, those of other groups also. To protect this system wars were fought. The strategic waters were brought under control of those who dominated the sea. The social structure by which trade could be controlled was defended in the belief that trade was the best or only way to maximize wealth and power, and for the elite wealth and power were conceived to be the supreme values of man.

The substitution of industrial production at home for overseas trade has made some of the assumptions questionable, but there is no doubt that it has completely altered the significance of most of the strategic geographic points made valuable by trade. The steam engine and its successors use fuel produced at fixed sites. Such fuel, produced at a cost (although such cost be far below the resultant energy available) placed a penalty on trade at a distance. It simultaneously permitted immense returns in production undertaken near the sources of energy. Hence the physical base of power is shifted from transportation as the primary factor controlling available power to industrial production. Gains formerly possible only through trade between those separated by distance are equalled and exceeded by means not requiring the costs of that social organization which is necessary to permit trade at a distance. With the advent of steam, stragic dominance of the sea and the flowing stream ceased to be means to monopoly of great power by an elite or a state.

There was thus set up alternative means to power within those states with access both to the sea and to the new fuel resources. The costs of imperialism, necessary to the overseas trader, were frequently merely an incubus to the industrialist who sought to exploit local resources and local markets. The monopoly on the generation of great power previously lying in the hands of the mercantile groups was brought into competition with the growing power of the industrial "isolationist." Moreover, the development of synthetic chemistry permitted the local production of substitutes for the old objects of foreign trade such as indigo, rubber, and silk. This reduced the dependency of major regions one upon another as did the development of metallurgy which provided new metals and alloys as substitutes for many minerals not locally produced. Elites must now compete for power, offering rewards for their support in the form of shared power, goods, or other values. Domestic suppliers can reward their followers more adequately and demand less sacrifice in return than their competitors who must maintain abroad, at great expense, social relations which are not necessary to the self-contained state. Such elites are, in so far as costs affect choice, likely to come into and remain in power. It is not certain then that in the future, industrial states will continue to be dominated by those who seek to preserve an elaborate system of interstate organization no longer necessary and in fact very costly to them.

The models for predicting technological change upon which a great deal of present action is based were developed from evidence provided by the events which took place in the West when industrial techniques were adopted. The value of the model as a means to predict either the conditions under which technological change is likely to take place or the consequences of that change is made doubtful by subsequent events other than those to which we have already alluded. When coal contended with sail as a means to increase the energy available to men, the technology by which the energy of coal was made available was very inefficient. In many cases the steam engine was little better than the converter which it sought to replace. Those who sought to exploit the energy of coal were not able to concentrate in their own hands power adequate ruthlessly to override those who sought to prevent industrialization. Hence compromise and persuasion were necessary means to change. In some areas today such compromise is not required. The borrowed technology permits the industrializer, by concentrating all the increased power gained from the limited application of technology in basic industry, rapidly to accumulate weapons and the means to mass communication and mass terror. He thus is able to wield power totally disproportionate to that available to those still dependent upon plants and animals as sources of energy with which to defend and maintain their way of life. Thus the assumption that future industrialization must be accompanied by that system of compromise which resulted in democracy in the West because such compromise took place in England or her colonies or in neighboring states requires for proof of present validity more than the resort to history. Such a system of compromise may in fact not appear, if at all, until there has been so extensive a development of industrialization as to result in territorial and functional specialization likely to produce competing factions among the industrial groups themselves.

It is also widely assumed in the West that the ideas and ideals which accompanied or were basic elements of the Protestant Revolt constitute the most efficient means to, and are thus reinforced by, the accumulation of converters of energy and the tools through which they may become productive. It is also regarded as being the most effective ideological base for the creation of the institutions necessary to the emergence and support in industrialism. Certainly history demonstrates that these concepts were, under the existing conditions, adequate to create great technological advance. But equally impressive is the fact that there are other very large areas of the world, equipped with adequate resources, capable of borrowing Western technology, who have refused to do so under the aegis of Protestant, Capitalist, Democratic ideas even in the face of the gains promised. We are thus confronted with the problem of determining whether or not another set of social arrangements and ideas, more compatible with the presently existing culture in those areas, or more ruthlessly capable of replacing that culture may prove to be a superior means to industrialization.

As we have indicated, both the major blocs now contending for world power assume that to secure adherence to the propositions which they seek to promote, changes in technology, particularly those which will result in a higher standard of living, must be forthcoming. It is of great concern then whether if one bloc fail to industrialize an area the competitor will be able to do so. It is altogether possible that either may fail in this endeavor and that effort to industrialize may in fact jeopardize rather than enhance the chance to gain an adherent by such effort. It will be necessary, in order to predict just what the likelihood of industrialization in a given area be, to discover just what technics are required for their operation.

As we have seen, empirical evidence indicates that there is considerable doubt that the historic model will serve to guide men in the future. The Marxian explanation of the way in which technological change will come about or what its consequences are likely to be has failed at every critical point to anticipate the changes which subsequently occurred and thus seems to be even less useful as a model for prediction.

Nevertheless, if it is to reveal to elites the alternatives that lie before them the Science of Peace must provide reliable means to anticipate the consequences of proposed changes, including particularly technological changes, since they are intimately connected with the power position of states.

We have already alluded to another one of the problems that confront us here, that of a reliable measure of alternatives. Various systems of transposing money costs into "real costs" have been developed. For some purposes they are a better measure of the technical efficiency of various social arrangements than is energy, which we have proposed. It is not assumed here that the calculation of energy costs will result in arrangements dictated solely by the effort to conserve energy. Such measurement is proposed rather in the effort to make it possible to compare the costs of two alternative sets of social arrangements. If energy costs of one be considerably in excess of the other, it is clear that he who utilizes the less costly means will be possessed of energy to secure other means beyond that which he would have had if he had utilized the more costly arrangements. We seek to permit the demonstration of what the logic of technics is, in terms which make possible comparison of the degree to which various social arrangements permit operations in accord with the logic of technics.

It is probable that in most cases less efficient arrangements than those maximizing return from energy in-put will actually prevail. Conscientious sabotage of what is technically feasible is a normal part of the operations of any society. To preserve or maximize other values technical efficiency is frequently sacrificed. But the fact remains and cannot be safely ignored that whoever, in the interests of preserving some other value, sacrifices technical efficiency, is handicapped in a test of force with an opponent utilizing energy more efficiently.

Present knowledge of the requirements of technology in purely empirical terms as discovered in *ad hoc* arrangements is enormous. Such arrangements, however, reflect past choices made among possible courses of action in terms of the values and social structure of those using them. Efficient use of technics has at times been furthered and at times hindered by prevalent social arrangements. The resultant technology for any time and place reflects those arrangements possible in technical terms—but only such of those as are socially sought.

It is not safe then to assume that under all conditions the only possible, or the most effective arrangements (in energy terms), are those which have arisen in some particular historic circumstance. The thesis that the only social arrangements which can accompany technics are those that have once been used and found to be successful may blind its proponents to effective alternatives. A competing elite might use them to such advantage as grossly to alter the existing arrangements which they sought to maintain. No amount of exhortation in the name of values not prevalent among an elite will serve to limit it to the use of only those alternatives considered to be legitimate by another elite. No amount of rationalization showing how their use of technology serves "higher" values will lessen the vulnerability of those who have in the interest of such values refused to make energy-wise efficient use of technology when confronted by a more powerful opponent whose very power may stem from the scorned arrangements.

If the Science of Peace is to reveal accurately to elites the consequences of their alternative courses of action, it will have to demonstrate such propositions as those to which we have just alluded. There are a great many areas in the world in which men have, in the interest of maintaining other values, refused to permit technical change. In others they have been unwilling to accept such change unless made possible by orthodox agencies which do not in fact produce change. Elites in such areas are confronted with the fact of immense power in the hands of their industrialized neighbors. In the absence of great force brought to their support by other states which have adapted their cultures to industrial technology, they are at the mercy of these powerful states. Nor may they depend upon the power of "International Law" to protect them. International law consists of that code which has the sanction of the power of states to enforce it, else it is nothing but a pious wish or the statement of non-demonstrable proposition. Emergent law will reflect the values of those who have the power to enforce it.

It is then of great significance to the Science of Peace that it be able to predict accurately the probable consequences of efforts to bring about technical change in the light of the known necessary requirements for such change.

If, for example, because of the character of technics the smallest efficient unit for the mechanical weaving of cloth be a loom producing at the rate of ten thousand yards per year, then no social production unit smaller than one making full use of that machine is efficient in conserving the use of energy. This single fact may require for the efficient use of technics tremendous alteration in the division of labor, the employment of children, or the aged in the home, and the functionality of other traditionally assigned roles in the community.

Similarly, if the smallest efficient open hearth furnace and attendant rolling mill produce ten thousand tons of steel a year, then no area in which consumers demand less steel is an efficient consumption unit for the use of the open hearth. This again may be of immense significance for the continuance of such social arrangements as local government, franchises, property, handcraft organizations, etc.

The introduction of industrialized agriculture may require relocation of much of the population, change the nature of claims on products and their functionality, destroy established markets and governments, change the nature of the family, and force changes in multitude of other relationships. A change in the means of production cannot then be regarded as a change only in such means; it must also be seen to have the effect of altering the means to other ends, for the means of production include behavior regarded as being an end in itself by those who will be disturbed by the necessity to abandon one way of life in favor of another.

It is frequently assumed that if technological change be proposed under the proper auspices, existing values will not be sacrificed. It is probable that some means to technological change are much less disruptive than others, and that some means involve the sacrifice of the values of different groups more than do others. The complaint of the Communist and many others opposed to certain aspects of "capitalism" is that technological change is made at the expense of workers, while capitalists reap the benefit of change. On the other hand some capitalist theorists assume that if governments abstain from interference, change introduced to secure profit *must* have the effect of benefiting the consumer. Obviously in both cases it is assumed that the cost of change is to be borne by some group and the benefits gained by another—"justice" consisting of benefits to that group whose values the critic felt to be the more significant. However it may become clear from research that in many cases effort to secure the adoption of new technology requiring some particular changes will be opposed no matter who the innovator or by what means he rationalizes his effort. Men may seek to preserve the values derivative from the self-sufficient small group even at the price of maintaining a less efficient technology. They may seek a way of life completely incompatible with those social arrangements necessary to the operation of technics making efficient use of the energy of coal, oil, and falling water. As we have noted, the history of technological change is in fact filled with evidence that coercion has been required to effect such change. In some cases intervention was direct and positive as in the Soviet Union, where a whole property system was destroyed to permit establishment of what was regarded as technologically more efficient. In other areas an indirect approach was made through subsidy, tariff, taxation, the use of "eminent domain" by "private industry," or the outright gift of valuable natural resources conditional upon the creation of particular technology. Sometimes the activity of the state was such as to deprive groups of the opportunity to react to change in a manner which would have interfered with, modified, or prevented change. The use of the Fourteenth Amendment to the Constitution of the United States is a case in point here, as is the use of injunction and a host of other legal devices.

Research may establish the fact then that in some areas technological change will take place, if at all, only where there is considerable coercion exercised by a minority supported by "foreign interests" against the values of the greatest part of a local population. Great efforts to increase public health in India, supported mostly by other motives than private profit to Englishmen, have backfired against the British there. The immense system of railroads, dams, and other public works made possible by British intervention have made the British extremely vulnerable to attacks by nationalists who complain that their "native culture" is being destroyed. Undoubtedly much of the action of the Soviet Union in attempting to introduce industry into the various national states of the Union has met with similar reaction. All proposals designed to change technology in the interest of some set of common supreme values presumably shared by all humanity must be critically examined to see whether in fact, in the area into which it is proposed to introduce them, they will have the desired effect. It may be that they will disturb or destroy values and social structure in that area felt to be more significant and more to be pursued than is the enhancement of those values which may be served by the new technology only at their expense.

As already indicated, a terrific amount of effort is currently being expended on both sides of the Iron Curtain on the assumption that to spread or to stop "Communism" it be necessary to change the technology of most of the world so that it may achieve a "Higher Standard of Living." Aside from the very questionable ability to change the technology of many areas of the world by the methods proposed, there is even more question as to the validity of the assumption that such change will result in new patterns of power and values more likely to resemble those being sought on either side than those now prevalent. For the West, for example, there is a very significant question whether or not the technological changes sought can be produced by any other method than one which will bring into being an industrial elite possessed of overwhelming power as compared with other members of that society, whose existence and growth in power are likely most adequately to be served by those very totalitarian methods which the Westerners profess to be attempting to prevent. For the U.S.S.R. there is a real question whether or not the establishment of new foci of power may not threaten the interests of the Soviet Fatherland by creating "Fascist" national states. With accurate knowledge of the minimum requirements of technics it may be possible to secure some desired change with less strain, if ingenuity be utilized to make technics serve the values of those in the "invaded" culture rather than those values which technology has hitherto, in the place of origin, been adapted to serve. It may in fact become apparent that adherence to one or the other bloc of the now contending states will depend upon the degree to which power is used to adapt technology to the local needs rather than used to cause the local population to adapt themselves to the peculiar use of technology found in its place of origin.

There is presently not much research designed to explore the opportunity to make modern technology serve the values of the "older" areas. The great bulk of present research stems from the efforts to solve certain immediate problems which confront those who use the technology to serve values of the system of which they are a part. Alternative uses of that technology would probably occur to those who were familiar with the needs and purposes of other people, but who are not now put in a position to discover emergent technology. The beginning of such research is to be found in some missionary experimental schools, in the export branches of some of the larger corporations, and apparently also some has been inspired by programs like the "Point Four" effort. Such research is, however, miniscule as compared with that devoted to the idea that technology can be transplanted to another area and made to create and to maintain whatever values it was originally created to serve.

Knowledge of the requirements of technology, and some insight into the values it is capable of creating and serving, will permit more accurate prediction of the results of efforts to change present values and social structure. Given knowledge of the basis of the power position of the present elite, the probability that it will seek to block such technological change, or that it will be successful in so doing, may also more accurately be predicted. The rise of a new elite to challenge the power of the old, based upon exploitation of the new technology, is also an alternative which must be explored if the Science of Peace is accurately to anticipate the alternative courses likely to be followed by a state.

We have already alluded to the fact that the centripetal effect of using coal, oil, and falling water as fuel may result in a situation in which the industrial elite in major power areas may seek to concentrate the gains possible from the use of new technology to a limited few in a limited area and will seek to avoid efforts on the part of others in other areas to get them to "invest" or otherwise provide the means for technological change in those areas. The history of the United States of America demonstrates that only occasionally, faced with threat from abroad, has there been any great effort, political or economic, to industrialize other than immediately adjacent areas. The rise of National Socialism in Germany seemingly represents a similar effort to concentrate in the Fatherland the gains of industrialism, and Russian Nationalism as opposed to International Communism would seem to lead to a similar effort. Only England, of the great industrial powers, has followed a pattern of great dispersion of industrial facilities, and this seems to have been required there by peculiar historic and geographic conditions which preceded industrialization rather than by something arising from technics *per se.*

Knowledge of modern technology and the institutions and values to which it has given rise may demonstrate that it is more likely to result in regional autarchy than in imperialism, cosmopolitanism, federal internationalism, or some other variety of interstate relationship. We have seen how many of the conditions which made industrialism possible in the West have subsequently been altered. If we are to predict the future power of many areas, we must explore methods that will show how and where industrialization is likely to take place in the absence of the impetus arising from those areas which originally built their systems around the use of the sailing ship and the institutions which served it. Perhaps the best that we can do at the moment is to construct some crude index which may reveal whether, if not exactly how, industrialization is likely to go forward.

We may find such an index in the calculation of the ratio between increase in population and increases in the accumulation of the converters of the energy of coal, oil, falling water, and perhaps atomic energy. Productivity is conveniently measured in terms of product per man per day. If converters making use of forms of energy which are less costly to man than is food be accumulated faster than the number of men who depend upon the product of those converters, the average product per man-day increases. If, however, men increase faster than these converters, production per man-day will fall to the level of that possible to men using food as fuel. Of two states, that one which more rapidly increases productivity per man-day will be in a position to use its increasing productivity either as direct threat to its neighbor in the form of physical coercion or, alternatively, to increase the material standard of living making that system more attractive to those who seek material well-being. A set of values and social structure which result in an increase in other converters, accompanied by a more rapid increase in men, is thus likely to place its users at a disadvantage as compared with those whose value system and structure permits increase in other converters at a much more rapid rate than the increase in men. Apparently in many areas of the world in the past, technological change has been stopped and even reversed by the demand that men be employed so that they can claim a functional right to eat and survive, in preference to the employment of machines which would deprive men of that right. Hence a population increasing rapidly, as compared with other converters, endangers the further development of technology and resultant power. The Science of Peace must then provide for the discovery on the one hand of all the factors which result in the increase of other converters than man, and on the other all those which result in an increase in man.

The Science of Peace
Deals with Capital

Empirically demonstrable propositions about the conditions under which it would be possible to accumulate converters is a tremendously significant concern for the Science of Peace.

There is considerable research devoted to the discovery of the means whereby capital may be accumulated. There is still, however, considerable disagreement even as to which concepts are most fruitful in this area. For example, the terrific growth of capital in the United States during that last twenty years, when almost every condition which was traditionally supposed to induce capital accumulation was ignored or repudiated, makes questionable the adequacy of traditional theory to account for what happened. Similar situations seem to have been encountered particularly in Russia and Germany. The empirical demonstration of theories showing the conditions necessary and favorable to the growth of capital under the conditions prevailing in various specific areas of the world provide the means to discover one important element in the index of productivity for that area.

There is, for instance, considerable evidence to be found in the efforts of the Kuomintang to industrialize within the framework of the Chinese community and family system. This should be carefully studied to see just why it never succeeded in breaking the hold of that traditional system. The same kind of research is of course possible and desirable in areas like India and Indonesia, the Philippines and Latin America. Even further analysis of the relation between the size of the capital base and the conditions which result in increases or decreases in the rate of accumulation in the U.S. should be enlightening, for the size of this base in relation to population in the U.S. transcends anything even dreamed of when most present theory was being first developed, and only in the United States can empirical evidence be found. What is suggested here is not that nothing has been or is being done but that those who seek to establish priorities for the Science of Peace should examine such research for lacunae and promote further research into these areas now not adequately being studied.

The Science of Peace
Deals with Population

Simultaneously with the study of the rate of accumulation of converters there should be increased study of the conditions that govern the growth of population. It is of important relevance to the Science of Peace to determine whether or not in some areas of the world, such as India, China, or Japan, it be possible ruthlessly to interfere with practices which result in population growth and at the same time to accelerate the process of capital accumulation. This is not to make a judgment that if such a system be possible it would be "moral" so to do. But it is necessary to see that if it be possible so to do, then some elite with power to act and with a set of values which sanction such acts (such as, for example, adherence to the proposition "the end justifies the means," "the inevitability of progress," or "the materialistic dialectic") may set out to accomplish this end by the necessary means with significant results for the future balance of power in the world.

There is abundant historic evidence of such attempts. There is much less adequate research dealing with empirical evidence of the conditions and values which specifically limit or encourage fertility and longevity. Instead there has been resort to pseudo-scientific extrapolation of some kind of mathematical series thought to represent "natural laws." More recently research has been extended to situational study of the values involved in begetting and rearing children, and in prolonging life. Controls over birth and death are apparently situational, and research must deal with specific relations in each of the areas in which it is sought to predict population growth.

Such study may permit an estimate as to whether it is to be so great in that area as to require either ruthless control by the industrialist or whether it will result in putting an effective stop to industrialization.

Westerners have long promoted longevity or "death control" while assuming that it is immoral to concern themselves with "birth control." It may now be required, if they seek to preserve their civilization, that they examine the consequences of fertility, even in the face of powerful forces which seek to stop efforts even to investigate the question as to why people have or refuse to have babies.

Research as to the relation between increase in population and other converters cannot be confined to nonindustrialized regions: it must also deal with empirically visible trends in the West. Perhaps the accurate construction of such an index is the most important single means to discover the prospects of future power for the various areas of the world.

Nor can research into other aspects of Western culture as it relates to technology be neglected on the assumption that there is to be found there adequate social structure to deal with putatively emergent events. There is certainly grave doubt that Western states themselves are prepared, while making use of modern technology, to continue to preserve their traditional values and social structure. Assuming that each of the national states of the West be content to accept a less technologically efficient set of social arrangements than is possible to it, in the interest of preserving other values, it is necessary to predict survival of those states, also to demonstrate that these arrangements are apparently adequate to resist the efforts of other states likely to seek forcibly to alter those arrangements.

Not only in the unindustrialized areas, but in much of the industrial world, the test for survival must be made in military terms.

The Science of Peace
Deals with War

The technology of modern warfare has altered tactical and strategic significance of weapons, terrain, man-power, location, technology, logistics, military organization, communications, and all other aspects of war. States that were once capable, through the effort of those who inhabited them, of defending themselves, have apparently ceased so to be; formerly strategic frontiers have become invitations to invasion; formerly inviolable strongholds have become the most vulnerable targets. An elite considering war as one of its alternatives must inform itself of the probable consequences of war in the light of all the conditions that now exist or can now be predicted. If the Science of Peace be useful in affecting decisions leading to peace or war, it also must assess these facts.

Even the answer to the question, "Who are those with means to resist sufficiently effectively and long enough to characterize the result as being war?" involves the calculation of the military power which one state or an elite can bring to bear on another. Trends in the development of aircraft and guided missiles, atomic weapons and other aspects of atomic energy, bacteriological and toxic weapons, new developments in the construction of submarines and other naval vessels, and the means to detect and destroy them—all become important facts for the Science of Peace. The study of Psychological Warfare, as contrasted with other means to the attainment of the values of the elite, must seek increasingly accurate knowledge of the science and art of propaganda, particularly through the use of mass media. This catalogue is not cited as inclusive. It is only suggestive of the fact that a great deal of information as to the putative costs and consequences of war is relevant to the Science of Peace, since only as such knowledge exists is it possible to calculate the relative cost and consequences of war as compared with other means to the acheivement of, or the costs of modification or abandonment of the pursuit of the values which may be dominant among those in a position to make war.

The consequences of new techniques of war-making seem to indicate that except for possibly one or two huge states, national states are no longer viable military units. This conclusion seemingly presents an insoluble dilemma for the West. Either it must permit the extension of the state far beyond the boundaries marking the extent of the nation, or it must succumb to powerful states not so limited. The past emphasis upon "National Self-Determination" by the democratic West is not fortuitous. In answer to the question, "How is the coercive power of the state to be made legitimate?" democratic theory said, "By limiting power of the state to areas and spheres where it is justified by moral and religious values locally shared by the population." The basic values— those regarded as so significant as to justify coercion to prescribe or proscribe what morals dictated should be so treated—were generally recognized as being co-extensive with the "Nation." Hence by democratic theory the legitimate area of a state coincided with boundaries drawn to correspond with the distribution of "nationals." In pursuit of such a system, national boundaries were drawn and "national sovereignty" established. In maintenance of such a system a tremendous portion of the productivity of the West, particularly of the United States, Germany, France, and Great Britain and her Commonwealth, has been expended. However sacred such values may be to those who share them, the facts demonstrate that to defend such a system of values there must exist and be implemented force adequate to resist successfully those who refuse to recognize the nation as the only legitimate unit for government and who seek to extend their power over large areas currently governed as small national states.

The survival of the national state system assumes that each national state can, with its own resources, meet the military test successfully. Otherwise a state is forced to choose either to sacrifice some of the values of the nation in order to secure the military cooperation of other national states, or to lose values of which it will be deprived if it be incapable of defending them by force.

It would be altogether surprising, in the light of what we have already said about the relationships between technics and values, if in every case the values which have in the past characterized the people of a nation should be those now necessary to permit it to develop the technology necessary for its own defense. It would be even more surprising if the materials and man-power necessary effectively to implement that technology should also be distributed in such manner as to permit each national state to defend itself against any probable opponent. Even a cursory examination of maps will demonstrate that there exists wide diversity of power among neighboring states. Weak states exert "sovereignty" only so long as it is convenient to the elite of those neighbors.

Research to establish how the extension of the national state will be undertaken may be based on a number of model combinations of states, each of which contains the raw materials, man-power, technology, industrial resources, and other factors necessary to make them viable in combat with other powers. It seems apparent at the moment that the U.S.A. and the U.S.S.R. are, in combat with other national states, capable of survival as presently organized political and military units.

Whether any other presently politically organized unit be so viable is apparently sufficiently doubtful that most other states seek some position vis-à-vis one of them, so as to assure that it will not be forced alone to defend itself and its interests against the other. However, it is possible, for example, that a combination of Western European States might combine territory and resources likely to be under their control during war—becoming sufficiently powerful to take a course of action dictated neither by fear of nor dependence upon the military might of the U.S.A. or the U.S.S.R. The Science of Peace, examining these potential states, may do much to predict which is likely to emerge as real by examining what sacrifice of which values of which groups of each set of nationals would be required to produce each of the model states. In large measure it can be predicted that the elite currently in power in a state will determine the values which will be sacrificed in order to maximize those by which the elite may retain its power and maximize its own values.

Alternate courses of action leaving intact the same "hard core" of basic values of the nation may involve the sacrifice of the values of different classes, regions, religions, nationalities, or other groups. Some combinations of states will be likely to emerge as a consequent of the existence in power of elites willing to sacrifice one set of values while the rise to power of another elite will lead to new combinations. Particularly in Western Europe, where democratic institutions permit rapid rotation of elites, the particular moment when states coalesce and create new structures, which more or less permanently secure the continued dominance of an elite, may have much to do with the future balance of power in the world. It is of great moment under whose auspices emergent states are born, and failure correctly to act at a given moment may have far reaching consequences. The changes necessary to secure some of these coalitions may be such as to require enormous sacrifice of the values of some groups. Whether or not in this situation the National State itself can, in order to secure the results demanded for cooperation with other states, be extended into those spheres previously immune to its action, becomes a moot question. Where military and technological necessity demand tremendous violation of the values hitherto held in some states to be inviolable, civil war may as frequently as international war threaten peace. In democratic states it may appear that no party which will require the sacrifice of values to assure the cooperation of those states necessary to secure military survival can maintain control over government through means considered constitutional. Yet failure to secure such cooperation may involve the sacrifice of even more of the basic morality of the nation, with little opportunity in any reasonable time to restore the conditions within which "constitutional" choice may again be made.

Some Conclusions

Throughout this work it is presumed that peace may be secured with approximately the same kind of human nature which now exists, as witness the periods of peace which alternate with the periods of war. For this reason there has been no consideration of the ways in which the various kinds of personality traits leading to conflict might be abated. It has been assumed that there is a great longing for peace, but much evidence also indicates that not all other values will be sacrificed to secure peace. For some people it is still "better to die on your feet than to live on your knees." The growing costs of war, however, even to the "victors," when war involves the Great Powers, and the tendency for each local dispute to become the occasion for the threat of a world war, make it more and more likely that honest investigation *will* demonstrate that alternatives to war *are* less costly for the antagonists than is war. We have earlier alluded to the fact that it seems probable that all of the conditions foreseen in Model V actually exist, except that condition required to make them manifest to elites considering war. If this can be established, then it is vitally important that it be established so as to convince the most skeptical of elites that it is true.

To this end research should be supported in such manner that the highest degree of intellectual integrity will be possible. It is also important that international forums exist in which research results can be made available to all interested men. The growth of independence on the part of the various branches of the United Nations, particularly UNESCO, will help to assure that it is not regarded by some groups as a mere sounding board for the dominant faction of those organizations.

Assuming the attainment of these goals, the Science of Peace may offer for the consideration of each elite the considered probability that through immigration, trade, propaganda, international organization, investment, or other means not involving force it can attain its goals. Alternatively it may reveal the staggering price of securing only some of them with *force.* Continued use of the military resources of the United Nations may result in making the putative cost of the use of force larger and larger, but in the foreseeable future most states will find it necessary to depend rather upon the strength of their allies than upon a powerful international "police force."

One of the more neglected areas of international research is that of demonstrating how elites may secure within their own borders many of the objectives they have been wont to seek outside. The glorification of internationalism as an end in itself has blinded many to these possibilities. But it is more and more possible by the use of metallurgy and synthetic chemistry for nations to secure within their presently legitimate limits many of the goods and services previously sought abroad. Failure to explore the possibilities here is another example of the way in which what was once defended only as a means has become an end in itself.

Emphasis has been laid upon empirical verification of all hypotheses used. We have sought to indicate that this will permit immediate demonstration of the use of science to guide policy. Such emphasis puts many of the proposed studies in the field of "Area Research." Empirical evidence that will establish even the most general propositions must be found in specific places. Throughout the work, we have stressed a "situational" estimate of the relations between values, social structure, demography, geography, and technology. It is only by dealing with particular people in particular places that we can discover whether or not generalizations are verifiable.

General theory based upon the truths found to be self-evident amongst men who are the product of one refined substratum of society has frequently been found to produce more distortion than crude empiricism itself. Whatever may be universal can be as easily established as true through use of evidence gathered at one place as another, and evidence gained by selecting for study those areas which are also useful for other purposes can in the meantime be valuable as a guide to more immediate policy. Action research has many drawbacks; but as the research program of such organizations as the Bureau of Agricultural Economics has demonstrated in the United States, it has simultaneously led to much more widespread support for "pure" science than it appears likely would otherwise have been forthcoming. As already cited, Clinical Medicine is another example of how the ability to demonstrate practical results has furthered the opportunity to do pure research. Those who are practically concerned with the problem, "Can there be a Science of Peace?" must take into account every act that will increase the probability of an affirmative answer, and action-oriented research seems to offer that kind of assurance.

As we have indicated, there is great variation in the kind and quality of research now available, and those who seek to provide usable answers will need to approach the problem of research support with an eclectic mind. Some needed elements of research are pouring out of laboratories, foundations, industries, and universities in such a flood that the job of synthesis is a much more important addendum than its field work or experiment. For other areas there is a pitifully small body of knowledge and almost no research resources. Here those interested in the Science of Peace should concentrate on the gathering of data, test promising but undemonstrated hypotheses, and provide for the recruitment and training of experts.

Many kinds of needed research will never get done by the research agencies now at work or likely to be founded. The Foundations will contribute to the growth of the Science of Peace if they make such research possible. Universities supported by governments or private bodies whose beliefs are likely to be shaken by research findings are unlikely to endanger their future support by revealing unpalatable truth —particularly about the "sacred" in their own culture. Governments dependent upon appropriations derived from men who must frequently be elected on the basis of what amounts to a popularity contest are not good prospects for the investigation of the scientific adequacy of those premises held dear by the populace. Parties and churches which already have in their possession the sacred books which reveal ultimate truth are not the most likely sponsors of uninhibited examination of the empirically verifiable. Even with the most complete certainty that their economic well-being will not be endangered by coming up with the "wrong" answer, it will be difficult to secure men willing to loosen the bonds of loyalty to friends and ideas sufficiently to permit them honestly and fearlessly to seek the truth. In the absence of secure status, secure income, and surroundings in which they will find social support in their lonely task, it will be impossible to get men of the type required.

Which brings us to what is perhaps the most significant criterion which those who seek to advance the Science of Peace must use. The men who run foundations are amateurs in the fields over which they exercise control. They customarily are able to get the estimate of the intellectual peers of those under consideration by which they can select men and the projects which they seek to undertake. In many of the areas here proposed for research there are no acknowledged experts. The men who run foundations must judge research by the character of the man who seeks to pursue it, in terms of his devotion to scientific truth and his apparent ability to perform the operations he plans to undertake. Reliance on established institutions, research techniques and concepts, and on orthodox scholars may deny us that creative imagination through which alone peace may become possible.

Index

187